QA
99
E75
2011

Erickson, Martin J.,
1963-

Beautiful
 mathematics.

Beautiful Mathematics

© 2011 by the Mathematical Association of America, Inc.

Library of Congress Catalog Card Number 2011939398

Print edition ISBN: 978-0-88385-576-8
Electronic edition ISBN: 978-1-61444-509-8

Printed in the United States of America

Current Printing (last digit):
10 9 8 7 6 5 4 3 2

Beautiful Mathematics

Martin Erickson
Truman State University

Published and Distributed by
The Mathematical Association of America

SPECTRUM SERIES

The Spectrum Series of the Mathematical Association of America was so named to reflect its purpose: to publish a broad range of books including biographies, accessible expositions of old or new mathematical ideas, reprints and revisions of excellent out-of-print books, popular works, and other monographs of high interest that will appeal to a broad range of readers, including students and teachers of mathematics, mathematical amateurs, and researchers.

MAA Service Center
P.O. Box 91112
Washington, DC 20090-1112
800-331-1622   FAX 301-206-9789

To Rodman Doll,
who mentored me in mathematics
when I was a high school student

Preface

Why are numbers beautiful? It's like asking why is Beethoven's Ninth Symphony beautiful. If you don't see why, someone can't tell you. I know numbers are beautiful. If they aren't beautiful, nothing is.

PAUL ERDŐS (1913–1996)

This book is about beautiful mathematical concepts and creations.

Some people believe that mathematics is the language of nature, others that it is an abstract game with symbols and rules. Still others believe it is all calculations. Plato equated mathematics with "the good." My approach to mathematics is as an art form, like painting, sculpture, or music. While the artist works in a tangible medium, the mathematician works in a medium of numbers, shapes, and abstract patterns. In mathematics, as in art, there are constraints. The most stringent is that mathematical results must be true; others are conciseness and elegance. As with other arts, mathematical ideas have an esthetic appeal that can be appreciated by those with the willingness to investigate.

I hope that this book will inspire readers with the beauty of mathematics. I present mathematical topics in the categories of words, images, formulas, theorems, proofs, solutions, and unsolved problems. We go from complex numbers to arithmetic progressions, from Alcuin's sequence to the zeta function, and from hypercubes to infinity squared.

Who should read this book? I believe that there is something new in it for any mathematically-minded person. I especially recommend it to high school and college students, as they need motivation, to study mathematics and beauty is a strong motivation; and to professional mathematicians, because we always need fresh examples of mathematical beauty to pass along to others. Within each chapter, the topics require progressively more prerequisite knowledge. Topics that may be too advanced for a beginning reader will become more accessible as the reader progresses in mathematical study. An appendix gives background definitions and theorems, while another gives challenging exercises, with solutions, to help the reader learn more.

Thanks to the people who have kindly provided suggestions concerning this book: Roland Bacher, Donald Bindner, Robert Cacioppo, Robert Dobrow, Shalom Eliahou, Ravi Fernando, Suren Fernando, David Garth, Joe Hemmeter, Daniel Jordan, Ken Price, Khang Tran, Vincent Vatter, and Anthony Vazzana. Thanks also to the people affiliated with publishing at the Mathematical Association of America, including Gerald Alexanderson, Don Albers, Carol Baxter, Rebecca Elmo, Frank Farris, Beverly Ruedi, and the anonymous readers, for their help in making this book a reality.

Contents

1
Imaginative Words

It is impossible to be a mathematician without being a poet in soul.

SOFIA KOVALEVSKAYA (1850–1891)

The objects of mathematics can have fascinating names. Mathematical words describe numbers, shapes, and logical concepts. Some are ordinary words adapted for a specific purpose, such as cardinal, cube, group, face, field, ring, and tree. Others are unusual, like cosecant, holomorphism, octodecillion, polyhedron, and pseudoprime. Some sound peculiar— deleted comb space, harmonic map, supremum norm, twisted sphere bundle, to name a few. Mathematical words have appeared in poems (see [19]). Let us look at some mathematical words.

1.1 Lemniscate

Consider the *lemniscate*, a curve shaped like a figure-eight[1] as shown in Figure 1.1. We learn in [46] that it gets its name from the Greek word *lemniskos*, a ribbon used for fastening a garland on one's head, derived from the island Lemnos where they were worn. By a coincidence, the end of the word lemniscate sounds like "skate," and one can (with practice and skill) skate a figure-eight. Skating a lemniscate is portrayed in the animated *Schoolhouse Rock* segment "Figure Eight," with the theme sung by jazz vocalist Blossom Dearie (1926–2009). She sings that a figure-eight is "double four," which is probably the Indo-European origin of the word "eight." In the animation, a girl daydreams of skating a figure-eight that turns into the infinity symbol ∞.

Figure 1.1. A lemniscate.

---

[1]Another curve, known as the Eight Curve, is perhaps closer to a figure-eight, but we will stick with the lemniscate because it is so graceful.

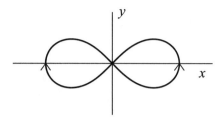

Figure 1.2. A lemniscate graph.

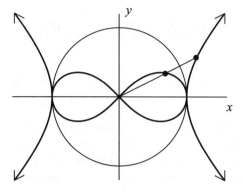

Figure 1.3. A lemniscate and a hyperbola as circular inverses.

As in Figure 1.2, a lemniscate can be graphed on an xy coordinate system by the parametric equations

$$x = \frac{\cos t}{1 + \sin^2 t}$$

$$y = \frac{\sin t \cos t}{1 + \sin^2 t}, \quad -\infty < t < \infty.$$

As the parameter t moves along the real number line, a point (x, y) in the plane traces the lemniscate over and over, following the right lobe counterclockwise and the left lobe in the clockwise direction.

Where do the parametric equations for the lemniscate come from? One way to obtain a lemniscate is by circular inversion of a hyperbola. In Figure 1.3, we see the hyperbola

$$x^2 - y^2 = 1$$

and the lemniscate as inverses with respect to the unit circle centered at the origin. Each point on the hyperbola is joined by a line segment to the origin. The point where it crosses the lemniscate is indicated. The length of the segment from the origin to the lemniscate is the reciprocal of the length of the segment from the origin to the hyperbola. The self-intersection point of the lemniscate corresponds to a point at infinity on the hyperbola.

The distance from the origin to a point (x, y) is $\sqrt{x^2 + y^2}$. To find the point on the lemniscate corresponding to (x, y) on the hyperbola, we make the transformation

$$(x, y) \mapsto \left(\frac{x}{x^2 + y^2}, \frac{y}{x^2 + y^2} \right).$$

This transforms the equation for the hyperbola into an equation for the lemniscate:

$$(x^2 + y^2)^2 = x^2 - y^2.$$

Starting with parametric equations for the hyperbola,

$$x = \sec t$$

$$y = \tan t, \quad -\infty < t < \infty,$$

we obtain, upon making the same transformation, the parametric equations for the lemniscate.

1.2   Centillion

What is the largest number you can name? A million is 10^3 thousand. A billion is 10^6 thousand. A trillion is 10^9 thousand. The largest number given in a dictionary list of numbers is typically a *centillion*, which is 1 followed by 100 groups of three zeros followed by another group of three zeros, or 10^{300} thousand, or

$$10^{303}.$$

A centillion is much larger than a *googol*, a coined term for 10^{100}, but much smaller than a *googolplex*, defined as 1 followed by a googol of zeros.

If you have ten dollars and seven cents in pennies then you have, in a way, a centillion. The number of ways of selecting a subset of the pennies is 2^{1007}, which is about 1.4 centillion.

1.3   Golden Ratio

Figure 1.4 shows a *golden rectangle*. If we remove the square on the shorter side, the remaining rectangle has the same proportions as the original rectangle. The *golden ratio* is

$$\frac{y}{x} = \frac{x}{y - x}.$$

This is

$$\frac{y}{x} = \frac{1}{\frac{y}{x} - 1},$$

or

$$\left(\frac{y}{x}\right)^2 - \frac{y}{x} = 1.$$

Completing the square,

$$\left(\frac{y}{x} - \frac{1}{2}\right)^2 = \frac{5}{4},$$

so

$$\frac{y}{x} - \frac{1}{2} = \pm\frac{\sqrt{5}}{2},$$

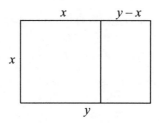

Figure 1.4. The golden rectangle.

and hence

$$\frac{y}{x} = \frac{1 \pm \sqrt{5}}{2}.$$

Since y/x is greater than 1, the positive sign applies. Therefore, the golden ratio, denoted by ϕ, is

$$\phi = \frac{1 + \sqrt{5}}{2} \doteq 1.6.$$

A *rational number* is a ratio of two integers, such as $4/7$ or $2/1$. An *irrational number* is a real number that isn't rational. The golden ratio is an irrational number. If ϕ were rational, then we could write $\phi = y/x$, where x and y are positive integers. From Figure 1.4, we see that ϕ also equals $x/(y - x)$. This is a representation of ϕ as a ratio of a pair of smaller positive integers. We could repeat the process, representing ϕ as ratios of smaller and smaller pairs of positive integers. But this would imply an infinite decreasing chain of positive integers, which is impossible. Therefore ϕ is irrational.

In Euclidean geometry, we can construct a line through any two points, a circle with any center and passing through any given point, and the intersection of two given lines, a line and a circle, or two circles. Figure 1.5 shows a construction of the golden rectangle using four lines and six circles.

For more about the golden ratio, see the delightful book [52].

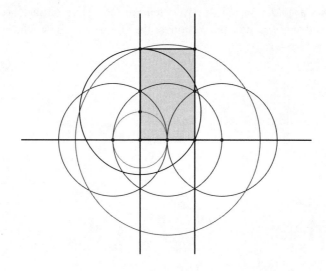

Figure 1.5. Construction of the golden rectangle.

Figure 1.6. Borromean rings.

1.4 Borromean Rings

Figure 1.6 shows three interlocking rings called *Borromean rings*, named after the Borromeo family in Italy whose coat of arms depicted them. No two rings are linked but if we remove one of them then the other two come apart.

Borromean rings exist as an abstract concept, but they do not exist in reality. The rings cannot be represented by three circles in 3-dimensional Euclidean space, even with arbitrary radii. The problem is that one cannot make rigid circles pass over and under each other in the required way. For a proof, see [31]. However, the sculptor John Robinson has shown that the interlocking configuration can be made with three squares or equilateral triangles instead of circles.

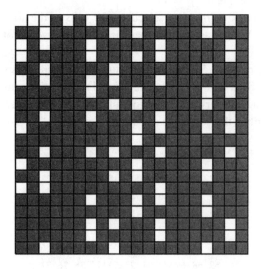

Figure 1.7. Sieve of Eratosthenes.

1.5 Sieve of Eratosthenes

A *prime number* is an integer greater than 1 with no positive divisors other than 1 and itself. For example, 13 is a prime number, but 10 is not because it is divisible by 2 and 5. There are infinitely many prime numbers, as proved in Euclid's *Elements*. Every integer greater than 1 is the product of primes in a unique way, the Fundamental Theorem of Arithmetic.

The *Sieve of Eratosthenes*, invented by Eratosthenes of Cyrene (c. 276–195 BCE), is an algorithm for listing all the primes up to a given number. The method is to eliminate proper multiples of known primes. Figure 1.7 shows the result of the Sieve of Eratosthenes on the numbers 2 through 400.

In the figure, the boxes represent the integers 2 through 400, reading left-to-right and top-to-bottom. Unshaded squares represent prime numbers and shaded squares represent composite (non-prime) numbers. The algorithm starts with the boxes unshaded. Then proper multiples of the first prime (2) are shaded (since they are divisible by 2 and hence composite). The next remaining number in order, 3, is a prime, and its proper multiples are shaded. The next remaining number is the prime 5, and its proper multiples are shaded. This continues for all primes up to 19 (those in the first row of the array). We need to sift out multiples of the primes 2, 3, 5, 7, 11, 13, 17, and 19, since 19 is the largest prime whose square is less than 400. Any composite number up to 400 must be divisible by one of them.

1.6   Transversal of Primes

Let p be a prime number. In a $p \times p$ square array consisting of the numbers 1 through p^2 (in left-to-right, top-to-bottom order), is there always a collection of p primes with no two of them in the same row or column? The solutions for $p = 2, 3,$ and 5 are unique. Figure 1.8 shows an example for $p = 11$.

A *transversal* of an $n \times n$ array is a selection of n cells of the array with no two in the same row or column. We are asking whether there is a transversal of primes in a $p \times p$ array.

Adrien-Marie Legendre (1752–1833) conjectured that there exists at least one prime number between consecutive squares N^2 and $(N + 1)^2$. This conjecture is still open. Legendre's conjecture is a necessary condition for our problem, since there must be a prime in the last row of the grid. Is the answer to our question possibly "no" for some prime p?

1	2	3	4	5	6	7	8	9	10	11
12	13	14	15	16	17	18	19	20	21	22
23	24	25	26	27	28	29	30	31	32	33
34	35	36	37	38	39	40	41	42	43	44
45	46	47	48	49	50	51	52	53	54	55
56	57	58	59	60	61	62	63	64	65	66
67	68	69	70	71	72	73	74	75	76	77
78	79	80	81	82	83	84	85	86	87	88
89	90	91	92	93	94	95	96	97	98	99
100	101	102	103	104	105	106	107	108	109	110
111	112	113	114	115	116	117	118	119	120	121

Figure 1.8. A transversal of primes.

1.7 Waterfall of Primes

Primes greater than 2 are odd and therefore upon division by 4 leave a remainder of 1 or 3. For example, $11 = 4 \cdot 2 + 3$ and $13 = 4 \cdot 3 + 1$. Among the first 1000 odd primes, 495 are of the form $4n + 1$ and 505 are of the form $4n + 3$. Thus, there are about half of each type. As the sequence of primes goes on, the distribution of primes into the two types is closer and closer to half-half. The waterfall of primes in Figure 1.9 depicts the way that prime numbers fall into the two classes, primes of the form $4n + 1$ on the right and primes of the form $4n + 3$ on the left. As the waterfall continues for all eternity, the difference between the number of primes of the two forms changes sign infinitely often.

Primes of different forms have different properties. For example, an odd prime is the sum of two squares of integers if and only if it is of the form $4n + 1$.

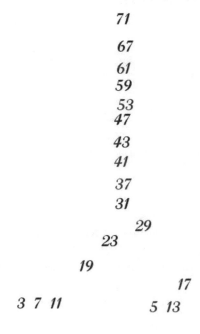

Figure 1.9. A waterfall of primes.

1.8 Squares, Triangular Numbers, and Cubes

Number theory is the study of properties of the counting numbers, $1, 2, 3, \ldots$.

A theorem of Joseph-Louis Lagrange (1736–1813) says that every positive integer is equal to the sum of four *squares* of integers. For example,

$$132 = 9^2 + 7^2 + 1^2 + 1^2.$$

A similar theorem, due to Carl Friedrich Gauss (1777–1855), asserts that every positive integer is equal to the sum of at most three triangular numbers. A *triangular number* is a number of the form $1 + 2 + \cdots + k$, for some positive integer k. So $10 = 1 + 2 + 3 + 4$ is a triangular number. The reason for this term is that dots representing the numbers 1 through

k can be stacked in the shape of a triangle. An example of Gauss's theorem is

$$100 = 91 + 6 + 3.$$

A third theorem of number theory, due to Pierre de Fermat (1601–1665), says that a *cube* (a positive integer of the form n^3) is never equal to the sum of two cubes. For instance, there is no way to write $10^3 = 1000$ as the sum of two cubes. This result is part of a more general assertion known as Fermat's Last Theorem, which was proved by Andrew Wiles in 1995. It says there are no positive integer solutions to the equation $x^n + y^n = z^n$, where n is an integer greater than 2.

Figure 1.10 represents these three theorems pictorially. In his diary, Gauss wrote an equivalent of the second equation accompanied by the exclamation *Eureka!*

A good reference on number theory is [37].

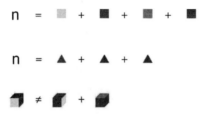

Figure 1.10. Three theorems of number theory.

1.9 Determinant

A *determinant* is an algebraic quantity that determines whether or not a system of linear equations has a solution.

Perhaps you are familiar with the formula for 2×2 determinants:

$$\begin{vmatrix} a & b \\ c & d \end{vmatrix} = ad - bc.$$

Did you know that the determinant is the area of a parallelogram? In Figure 1.11, the area of the gray parallelogram, spanned by the vectors (a, b) and (c, d), is the area of the rectangle minus the areas of two triangles and two trapezoids:

$$(a + c)(b + d) - \frac{1}{2}ab - \frac{1}{2}cd - \frac{1}{2}c(b + b + d) - \frac{1}{2}b(c + a + c)$$

$$= ad - bc.$$

In any dimension, a determinant is equal to the signed volume of the parallelepiped spanned by its row vectors.

1.10 Complex Plane

Complex numbers were treated with skepticism when they were first introduced in the 1500s. What sense can be made of the number $\sqrt{-1}$?

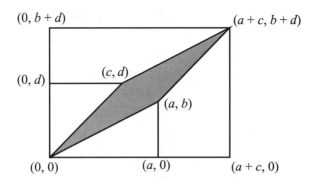

Figure 1.11. A 2×2 determinant as an area.

Since the square of a positive number is positive, and the square of a negative number is positive, and zero squared is zero, it appears that the square of no number can be -1. However, this is the definition of i, the imaginary unit.

The equation $x^2 + 1 = 0$ has no solution in the field of real numbers $\mathbf{R}$. But it is possible to solve it in a field that contains $\mathbf{R}$. The field is created by adding a new element, i, to $\mathbf{R}$ that has the property that $i^2 + 1 = 0$. Once i is added, other numbers must be added in order to ensure that the structure is a field. We call this field the field of complex numbers, $\mathbf{C}$. In this field, every polynomial equation with complex coefficients has a complex root.

The field $\mathbf{C}$ can be identified with the plane $\mathbf{R}^2$ in a natural way. However, the two structures are different. Multiplication of two vectors in $\mathbf{R}^2$ cannot be defined to produce a new vector so as to form a field. (The dot product of two vectors produces a number, not a vector. The cross product is defined for two vectors in $\mathbf{R}^3$, but this multiplication does not produce a field, since it isn't commutative.) However, a vector multiplication is possible in $\mathbf{C}$.

The main reason for defining this new field is the realization of these two properties: algebraic closure and existence of a multiplication.

In the construction identifying $\mathbf{C}$ and $\mathbf{R}^2$, we identify each real number $r \in \mathbf{R}$ with the ordered pair $(r, 0)$, we identify i with the ordered pair $(0, 1)$, and we identify the number $a + bi$ with the ordered pair (a, b). This construction was first carried out by the Norwegian mathematician Caspar Wessel (1745–1818). For an engaging account of the early history of complex numbers, see [36].

Once we have defined the element i such that the equation $i^2 = -1$ holds, we have defined the field of complex numbers $\mathbf{C}$. The relation $i^2 = -1$ induces a rotational product for the whole complex plane. Hence, $(\mathbf{R}^2, +, \cdot)$ forms a field where addition is ordinary vector addition. We call $(\mathbf{R}^2, +, \cdot)$ the *complex plane*, and the points $(a, b) \in \mathbf{R}^2$ are called *complex numbers*. They are ordinary points in the ordinary plane, but with a way to multiply them to get another point. Thus, the plane with vector addition and this rotational product is a field. See Figure 1.12.

Many polynomials don't have real zeroes, for example, $x^2 + 1$. The complex numbers are built upon the reals and a zero of this polynomial. The nontrivial fact that every polynomial has a real zero was first proved by Carl Friedrich Gauss (1777–1850) in the early 1800s, and is called the Fundamental Theorem of Algebra.

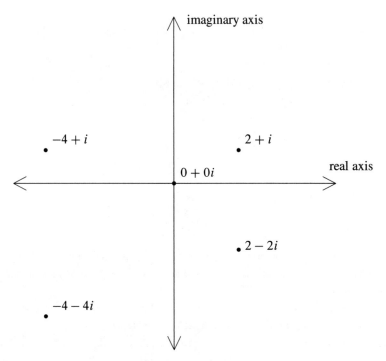

Figure 1.12. The complex plane.

We define the sum of two complex numbers by

$$(a + bi) + (c + di) = (a + c) + (b + d)i,$$

or, in terms of ordered pairs,

$$(a, b) + (c, d) = (a + c, b + d).$$

Multiplication is defined based on the rule that $i^2 = -1$, so that

$$(a + bi)(c + di) = (ac - bd) + (ad + bc)i,$$

or, in terms of ordered pairs,

$$(a, b) \cdot (c, d) = (ac - bd, ad + bc).$$

The formula for multiplication can be remembered by writing (a, b) as $a + bi$ and using ordinary multiplication of binomials, replacing i^2 by -1 wherever it occurs.

Since $(a, 0) + (b, 0) = (a + b, 0)$ and $(a, 0) \cdot (b, 0) = (ab, 0)$, we can identify the x-axis with the real line. We call $(x, 0)$ a *real number*.

As $(0, 1) \cdot (0, 1) = (-1, 0)$, we can identify $(0, 1)$ with i. This is also denoted by i and the y-axis is called the *imaginary axis*. We call $(0, y)$ a *pure imaginary number*.

We see that $(b, 0) \cdot (0, 1) = (0, b)$ and thus $(a, b) = (a, 0) + (0, b) = (a, 0) + (b, 0) \cdot (0, 1)$. That is, $(a, b) = a + bi$ where a and b are real numbers called the *real* and *imaginary* parts of the complex number $a + bi$. We denote the real part of z by $\Re z$ and its imaginary part by $\Im z$.

We define $-(a + bi)$ to be the complex number corresponding to the ordered pair $(-a, -b)$. Also, we define the difference of $a + bi$ and $c + di$ to be $(a + bi) - (c + di) = (a - c) + (b - d)i$.

To prove that $\mathbf{C}$ is a field, we need to show that an arbitrary complex number $z = a + bi$ has a multiplicative inverse z^{-1}. We can do this by rationalizing the denominator:

$$z^{-1} = \frac{1}{a + bi} = \frac{a - bi}{(a + bi)(a - bi)} = \frac{a - bi}{a^2 + b^2} = \frac{a}{a^2 + b^2} - \frac{b}{a^2 + b^2}i.$$

To evaluate the complex quotient z_1/z_2 is easy; we compute $z_1 z_2^{-1}$.

We can show that complex multiplication is associative, commutative, distributive over addition, and has the unit $1 = 1 + 0i = (1, 0)$. This means that $(\mathbf{R}^2, +, \cdot)$ is a field, and we denote it by $\mathbf{C}$.

The *modulus* or *absolute value* of a complex number $a + bi$ is its distance, as a point in the plane, from the origin $(0, 0)$. That is, $|a + bi| = \sqrt{a^2 + b^2}$.

The distance between the complex numbers $z_1 = a_1 + b_1 i$ and $z_2 = a_2 + b_2 i$ is the usual Euclidean distance in the plane between (a_1, b_1) and (a_2, b_2), which is

$$\sqrt{(a_2 - a_1)^2 + (b_2 - b_1)^2}.$$

The *conjugate* of $z = a + bi$ is $\bar{z} = a - bi$. Geometrically, the vector z is reflected in the x-axis to produce the vector $\bar{z}$.

Leonhard Euler (1707–1783) discovered a fundamental connection between the sine and cosine functions and the exponential function. When we consider them as functions of a real variable, there doesn't appear to be any connection among them. Sine and cosine are bounded, periodic and take on negative values, which contrasts with the behavior of the exponential function, which has none of these properties. To see the relationship, we look at the exponential function with a purely imaginary argument, or, more precisely, determine how the exponential function should be extended to the y-axis of $\mathbf{R}^2$. The functions have power series expansions

$$\cos \theta = 1 - \frac{\theta^2}{2!} + \frac{\theta^4}{4!} - \frac{\theta^6}{6!} + \cdots$$

$$\sin \theta = \theta - \frac{\theta^3}{3!} + \frac{\theta^5}{5!} - \frac{\theta^7}{7!} + \cdots$$

$$e^x = 1 + \frac{x}{1!} + \frac{x^2}{2!} + \frac{x^3}{3!} + \frac{x^4}{4!} + \frac{x^5}{5!} + \frac{x^6}{6!} + \frac{x^7}{7!} + \cdots.$$

Euler considered $e^{i\theta}$. Because $i^{4k+2} = -1$ and $i^{4k} = 1$, the coefficients of even powers of θ in the expansions of $e^{i\theta}$ and $\cos \theta$ are the same. And because $i^{4k+1} = i$ and $i^{4k+3} = -i$, the coefficients of odd powers of θ in the expansion of $e^{i\theta}$ and $i \sin \theta$ are the same. Thus, we have Euler's formula

$$e^{i\theta} = \cos \theta + i \sin \theta.$$

This is not really a derivation since we haven't defined the cosine, sine, or exponential functions with complex arguments, much less determined their power series. So, actually, Euler's formula is an insight into what becomes the definition of the exponential function with a pure imaginary argument.

Using Euler's formula, we can write a complex number $z = a + bi$ in polar coordinates form as $z = re^{i\theta}$. For example,

$$5 = 5e^{i0}, \quad -5 = 5e^{i\pi}, \quad i = e^{i(\pi/2)}, \quad -i = e^{i(-\pi/2)}, \quad -1 + i = \sqrt{2}e^{i(3\pi/4)}.$$

As a further example, the circle of radius 5 centered at z_0 is $|z - z_0| = 5$, or equivalently, $z = z_0 + 5e^{i\theta}$, where $0 \le \theta < 2\pi$.

We have seen that we can represent a complex number as an ordered pair of real numbers or in polar coordinates. We can also represent it as a *vector*

$$\begin{bmatrix} a \\ b \end{bmatrix},$$

where a and b are real numbers.

And we can represent a complex number as a 2×2 matrix:

$$\begin{bmatrix} a & -b \\ b & a \end{bmatrix} \quad \text{or} \quad r \begin{bmatrix} \cos\theta & -\sin\theta \\ \sin\theta & \cos\theta \end{bmatrix}.$$

The second matrix is a rotation matrix corresponding to counterclockwise rotation about the origin by the angle θ in radians.

Multiplication by a fixed matrix is a linear transformation of the plane. This allows us to apply complex numbers to problems of geometry. An example of the interplay between complex numbers and plane geometry is the description of the isometries of the Euclidean plane in terms of complex numbers. See Isometries of the Plane in Chapter 4.

2

Intriguing Images

Mathematics, rightly viewed, possesses not only truth, but supreme beauty—a beauty cold and austere, like that of sculpture, without appeal to any part of our weaker nature, without the gorgeous trappings of painting or music, yet sublimely pure, and capable of a stern perfection such as only the greatest art can show. The true spirit of delight, the exaltation, the sense of being more than Man, which is the touchstone of the highest excellence, is to be found in mathematics as surely as poetry.

—BERTRAND RUSSELL (1872–1970), *The Study of Mathematics*

Many mathematical concepts are embodied in diagrams, drawings, and other kinds of images. A sketch may illustrate a theorem. A picture may point the way to new mathematics. Let us look at some mathematical images and learn about the mathematics behind them.

2.1   Square Pyramidal Square Number

The equation

$$1^2 + 2^2 + 3^2 + \cdots + 24^2 = 70^2$$

might seem, at first glance, to be a miscellaneous mathematical fact, but it is special. Édouard Lucas (1842–1891) posed a problem, called the Cannonball Puzzle, which asked for a number N such that N cannonballs (spheres) can be placed in a square array, or in a pyramidal array with a square base. Thus, Lucas asked for a solution in integers[1] to the equation

$$1^2 + 2^2 + 3^2 + \cdots + m^2 = n^2,$$

where $N = n^2$. From the formula for the sum of the first consecutive m squares, the equation becomes

$$\frac{m(m + 1)(2m + 1)}{6} = n^2.$$

Lucas suspected, but was unable to prove, that the only solution is $m = 24$ and $n = 70$, with $N = 4900$, as above. That is, 4900 is the only *square pyramidal square number* greater

---

[1] A polynomial equation required to have integer solutions is called a *Diophantine equation*, after Diophantus of Alexandria (c. 200–c. 284 B.C.E.), whose book *Arithmetica* treats equations of this kind.

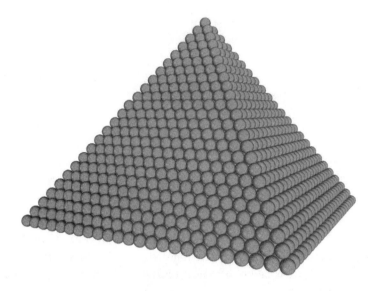

Figure 2.1. A pyramid of 4900 spheres.

than 1. Mathematicians after Lucas succeeded in proving this (see [4] for a particularly clear treatment). Figure 2.1 shows the pyramid of 4900 spheres.

The set of solutions to

$$\frac{x(x+1)(2x+1)}{6} = y^2$$

comprise what is called an *elliptic curve*. The study of elliptic curves is an active and fascinating area of mathematics. An excellent book is [53].

The solution to the Cannonball Puzzle is the basis for the existence of a famous lattice in 24 dimensions known as *Leech's lattice*, discovered by John Leech (1926–1992). A lattice is a regularly repeating pattern of points, such as the intersection points of graph paper. If we put a sphere of the same radius around each lattice point so that the spheres just touch, then the lattice yields a *sphere packing*. Leech's lattice has a remarkably high contact number (the number of nearest neighbors of each lattice point): 196,560. In the sphere packing associated with Leech's lattice, every sphere touches exactly 196,560 other spheres. This is the maximum contact number for a lattice sphere packing in 24 dimensions.

Leech's lattice can be constructed from a Lorentzian lattice in 26 dimensional space. The vector $w = (0, 1, 2, 3, \ldots, 24; 70)$ has length 0 in this space, since in the Lorentzian metric we compute length to be the square root of the sum of the squares of the coordinates with the exception that the last coordinate square is subtracted. Leech's lattice (a 24-dimensional linear space) is the quotient space $w^{\perp}/w$ of the orthogonal space to w (dimension 25) and the space spanned by w (dimension 1). The definitive reference on lattices and sphere packing is [14].

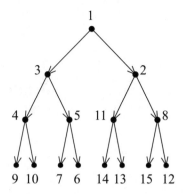

Figure 2.2. An order-preserving labeling of the full binary tree of order 4.

2.2   Binary Trees

Consider

$$\binom{2^n - 2}{2^{n-1} - 1}\binom{2^{n-1} - 2}{2^{n-2} - 1}^2\binom{2^{n-2} - 2}{2^{n-3} - 1}^{2^2} \cdots \binom{2}{1}^{2^{n-2}}, \quad n \geq 1.$$

The expression counts the number of order-preserving labelings of the *full binary tree* of order n, with the integers $1, \ldots, 2^n - 1$. The full binary tree of order n is a directed graph with a top node joined by arrows to two nodes at the next level down; each of these is joined by arrows to two nodes at the next level down, and so on for n levels in all. In an order-preserving labeling, each node is labeled with a smaller number than the labels of any of its descendents. Figure 2.2 shows an example with $n = 4$.

To see that this is what is counted, notice that the 1 must go at the top node of the tree. Then there is a choice of half of the remaining elements to go into the left subtree. It may be made in $\binom{2^n-2}{2^{n-1}-1}$ ways. This leaves the other elements in the right subtree. The least element in each subtree must go on top. Repeating, allowing for all the choices in all the branches at each level, gives the expression.

Call this number $f(n)$. The table below gives some numerical values.

n	$f(n)$
1	1
2	2
3	80
4	21964800
5	74836825861835980800000

If $n = 1$, the expression is an empty product which we define to be 1.

The labeled binary tree in Figure 2.2 is an example of a data structure in computer science called a *heap*. A heap can be viewed as a labeled subtree of a full binary tree. Figure 2.3 illustrates a heap on the set $\{1, 2, 3, 4, 5, 6, 7, 8, 9\}$. Heaps allow for quick insertion or deletion of minimum or maximum values, and they are used in a sorting algorithm called heapsort.

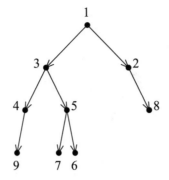

Figure 2.3. A heap.

2.3 Bulging Hyperspheres

Figure 2.4 shows a square of side length 4 circumscribing four circles of radius 1. By the Pythagorean theorem, the radius of the small circle in the middle of the larger circles is $\sqrt{2} - 1$. What happens if we generalize to any dimension $d \geq 1$? Suppose that we have a hypercube of side 4 in d-dimensional space, containing 2^d hyperspheres of radius 1. See Volume of a Ball in Chapter 3. We can place the hypercube so that its center is at the origin and the unit hyperspheres are centered at $(\pm 1, \pm 1, \ldots, \pm 1)$. By the Pythagorean theorem, the distance between the center of one of the unit hyperspheres and the hypersphere that sits in the middle of them is $r = \sqrt{d}$. Hence the radius of the small hypersphere is

$$r = \sqrt{d} - 1.$$

For $d = 4$, the radius of the small hypersphere is 1, the same as the radii of the other hyperspheres. For $d > 4$, the small hypersphere is larger than those that surround it. For $d = 9$, the radius is 2 and the small hypersphere touches the sides of the hypercube. For $d > 9$, the small hypersphere bulges outside the hypercube!

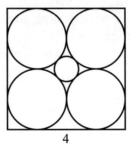

Figure 2.4. A circle surrounded by other circles.

2.4 Projective Plane

A *projective plane* is a geometry in which every two points determine a line and every two lines intersect in exactly one point. There are no parallel lines! Figure 2.5 shows a thirteen-point projective plane.

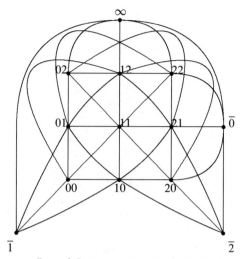

Figure 2.5. A projective plane of order three.

The thirteen-point projective plane has thirteen lines (in the figure, some of the lines are curved). Every line contains four points and every point is on four lines. Nine of the points are labeled with coordinates 00 through 22. The other four points, called *ideal points*, are labeled $\bar{0}, \bar{1}, \bar{2}$, and ∞.

We call this little universe a projective plane of order three (one less than the number of points per line). A *projective plane* of order n is a collection of $n^2 + n + 1$ points and $n^2 + n + 1$ lines such that each line contains $n + 1$ points, each point lies on $n + 1$ lines, every two points determine a unique line, and every two lines intersect in exactly one point. There exists a projective plane of order equal to any power of a prime number. No one knows if there is a projective plane whose order is not a prime power. The smallest integer greater than 1 that isn't a prime power is 6, and Gaston Tarry (1843–1913) proved that there is no projective plane of order 6. The next feasible order, 10, has been ruled out by a combination of mathematics and computer calculations: there is no projective plane of order 10. The existence of a projective plane of order 12 remains an open question. If it exists, it would have 157 points and 157 lines.

2.5   Two-Colored Graph

In graph theory, a *graph* is a collection of vertices and a collection of edges joining pairs of vertices. The edges may be straight or curved and may cross. Figure 2.6 shows a *complete graph* on 17 vertices. It is complete because every two vertices are joined by an edge.

The edges of the graph are colored with two colors, indicated by dark lines and light lines. The coloring has the property that there exist no four vertices all of whose six edge connections are the same color. However, every two-coloring of the edges of the complete graph on 18 vertices must have four vertices all of whose edge connections are the same color. This statement is an instance of a combinatorial result called Ramsey's theorem. Ramsey's theorem, discovered by Frank Ramsey (1903–1930), says that for every n, there exists a least integer $R(n)$ so that no matter how the edges of a complete graph on $R(n)$ vertices are two-colored, there exist n vertices all of whose edge connections are the same color. Thus $R(4) = 18$. Can you show that $R(3) = 6$?

The coloring of Figure 2.6 has a cyclic symmetry, with every vertex joined by dark edges to the vertices at steps 1, 2, 4, 8, 9, 13, 15, and 16 clockwise around the circle.

A good reference on graph theory is [54].

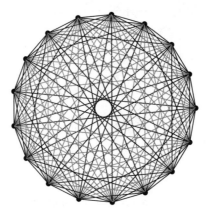

Figure 2.6. A two-coloring of the complete graph on 17 vertices.

2.6   Hypercube

To sketch a two-dimensional drawing of a cube, draw two squares separated by a little distance and draw four lines joining corresponding vertices. We can go a step further and draw a *hypercube*, a four-dimensional cube. Draw two cubes a little distance apart and draw lines joining corresponding vertices. The drawing has sixteen vertices and thirty-two edges, each vertex joined to four other vertices. Figure 2.7 shows one way to draw the picture.

A hypercube can be defined combinatorially as the set of sixteen binary strings of length four, e.g., 0110, where two strings are joined if and only if they differ in exactly one place. For example, the strings 0110 and 0111 are joined. This way of thinking about a hypercube is useful in constructing an example of a graph coloring. In the exercises in Appendix B, you are asked to give a three-coloring of the edges of a complete graph on 16 vertices such that there exists no triangle all of whose edges are the same color. In such a coloring, each single-color subgraph is a hypercube with the diagonals added.

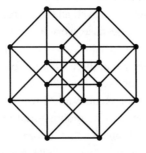

Figure 2.7. A hypercube.

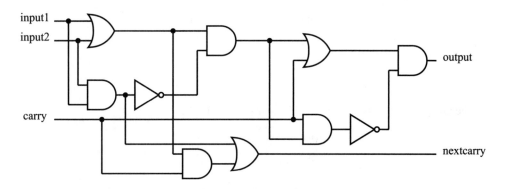

Figure 2.8. A full adder.

2.7 Full Adder

Figure 2.8 shows a logic diagram for an integrated circuit known as a *full adder*. It is the backbone of the arithmetic unit of a computer.

The circuit performs binary addition. Given two input bits (0 or 1) and a previous carry bit, the circuit adds the numbers in binary and yields an output bit and a next carry bit. Here is the *truth table* for this circuit.

input1	input2	carry	output	next carry
0	0	0	0	0
0	0	1	1	0
0	1	0	1	0
0	1	1	0	1
1	0	0	1	0
1	0	1	0	1
1	1	0	0	1
1	1	1	1	1

The elements in a full adder are called *logic gates*. The basic types are AND gates, OR gates, and NOT gates. They perform the logical operations described by their names. In combination they can create complex circuits such as the full adder.

The NOT gate is the simplest. It changes an input of 1 to 0, and 0 to 1.

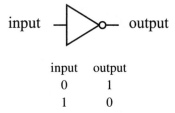

input	output
0	1
1	0

The AND gate yields an output of 1 if and only if both inputs are 1.

input2 ─┐
input1 ─┘ output

input1	input2	output
0	0	0
0	1	0
1	0	0
1	1	1

The OR gate yields an output of 1 if either input is 1.

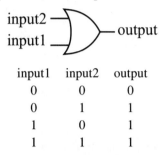

input1	input2	output
0	0	0
0	1	1
1	0	1
1	1	1

All aspects of a computer's thinking, including memory, are formed from these building blocks. The mathematical underpinnings are discussed in [43].

2.8   Sierpiński's Triangle

Figure 2.9 shows a *fractal* shape called *Sierpiński's triangle*, introduced by Wacław Sierpiński (1882–1969). The triangle is shown at the sixth step of its formation. Starting with a solid equilateral triangle, at the first step the triangle is divided into four equal-size equilateral triangles and the middle one is removed. At each subsequent step, this process is repeated on the remaining solid equilateral triangles.

The area of the Sierpiński triangle is 0. Suppose that the starting equilateral triangle has area 1. At the first step, the area is $3/4$ of the original area since one of the four sub-triangles is removed. At each further step, the area is reduced to $3/4$ of the previous area. Hence, the area at step k is

$$\left(\frac{3}{4}\right)^k,$$

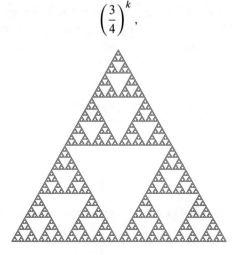

Figure 2.9. The Sierpiński triangle.

and this tends to 0 as k tends to infinity. Sierpiński's triangle isn't one-dimensional or two-dimensional. It has a fractional dimension, called *Hausdorff dimension*, equal to

$$\frac{\log 3}{\log 2} = 1.585\ldots.$$

The reason is that Sierpiński's triangle is the union of three copies of itself, each scaled down by a factor of two.

2.9   Squaring Map

Define a graph whose vertices are the integers modulo n. For two vertices A and B, draw a directed arrow from vertex A to vertex B if

$$A^2 \bmod n = B.$$

We call this graph the *squaring map modulo n*. You may want to draw the squaring map for some small values of n, such as $2 \leq n \leq 10$. Here is the squaring map modulo 25.

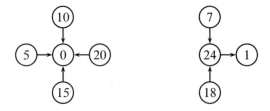

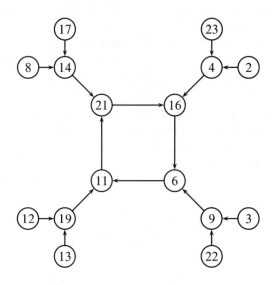

The arrows for the vertices that map to themselves, 0 and 1, aren't shown. There is a directed cycle of length four, namely,

$$6 \rightarrow 11 \rightarrow 21 \rightarrow 16 \rightarrow 6.$$

We call the sets $\{0\}$, $\{1\}$, and $\{6, 11, 21, 16\}$ *attractors*. If we start at a vertex and follow the arrows, we end up in an attractor.

If there is a directed path from one vertex to another, we say that the vertices are in the same *component* of the graph. The squaring map modulo 25 has three components.

The squaring map modulo n always has exactly one attractor in each component (exercise). Hence, the squaring map modulo n always has at least two components, corresponding to the attractors $\{0\}$ and $\{1\}$. It can be shown (more difficult) that the squaring map modulo n has exactly two components if and only if n is a power of 2 or a Fermat prime. A *Fermat prime* is a prime of the form $F_j = 2^{2^j} + 1$, where $j \geq 0$. The only known Fermat primes are $F_0 = 3$, $F_1 = 5$, $F_2 = 17$, $F_3 = 257$, and $F_4 = 65537$. What does the squaring map look like when $n = 17$? The graph appears, with a few minor differences, as a figure earlier in this chapter.

Using a computer, can you find the number of components and the size of a largest attractor when n is 1,000,000? For a complete solution to the squaring map problem, see [24].

2.10   Riemann Sphere

The Riemann sphere builds on the definition of the complex plane as a representation of complex numbers.

It is possible to model the real line and the plane by including a new point that we call ∞. The models are called the *extended* real line or plane. What do they look like? The extended real line is a circle, since $\pm\infty$ are the same point. Similarly, the extended plane is a sphere called the *Riemann sphere*, named after Bernhard Riemann (1826–1866); see Figure 2.10. Stereographic projection is a bijection between points in the plane and points on the punctured Riemann sphere (with the North Pole removed). The North Pole is identified with ∞.

We take the Riemann sphere to be a unit sphere centered at $(0, 0, 0)$. The North Pole is $(0, 0, 1)$ and the South Pole is $(0, 0, -1)$. In stereographic projection, a point $z = x + yi \equiv (x, y, 0)$ in the complex plane (which is equivalent to the xy-plane) is mapped to the point (x', y', z') on the sphere so that $(0, 0, 1)$, $(x, y, 0)$, and (x', y', z') are collinear. The origin is mapped to the South Pole; the unit circle in the plane is mapped to the equator; circles centered at the origin of radius greater than (respectively, less than) 1 are mapped to latitudinal circles in the northern (respectively, southern) hemisphere; and lines through the origin are mapped to longitudinal circles.

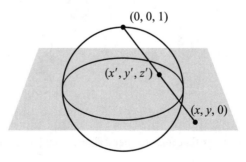

Figure 2.10. Stereographic projection.

We define the *extended complex plane* $\widehat{C}$ to be the complex plane **C** together with the point at infinity ∞; that is, $\widehat{C} = \mathbf{C} \cup \{\infty\}$. Stereographic projection is a correspondence between $\widehat{C}$ and the Riemann sphere.

Let's find the correspondence between $(x, y, 0)$ and (x', y', z'). The line determined by $(0, 0, 1)$ and $(x, y, 0)$ has parametric form

$$x' = \lambda x, \quad y' = \lambda y, \quad z' = 1 - \lambda, \quad \lambda \in \mathbf{R}.$$

Since (x', y', z') lies on the unit sphere,

$$(\lambda x)^2 + (\lambda y)^2 + (1 - \lambda)^2 = 1,$$

or

$$\lambda^2(x^2 + y^2 + 1) - 2\lambda = 0,$$

a quadratic equation in λ. One of the roots is $\lambda = 0$, which corresponds to the North Pole. The other is

$$\lambda = \frac{2}{x^2 + y^2 + 1} = \frac{2}{|z|^2 + 1}.$$

This yields

$$x' = \frac{2\Re z}{|z|^2 + 1}, \quad y' = \frac{2\Im z}{|z|^2 + 1}, \quad z' = \frac{|z|^2 - 1}{|z|^2 + 1}.$$

In the reverse direction, we obtain from the parametric formulas

$$x = \frac{x'}{1 - z'}, \quad y = \frac{y'}{1 - z'}.$$

Stereographic projection gives a correspondence between lines and circles in the plane and circles on the sphere. A curve that is either a line or a circle is called a "lircle." See Figure 2.11.

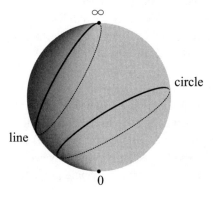

Figure 2.11. Lircles on the Riemann sphere.

Let's prove this. The equation for a lircle is

$$A(x^2 + y^2) + Bx + Cy + D = 0.$$

If $A = 0$, then we have a line. The condition that the circle is not degenerate is

$$4AD < B^2 + C^2.$$

We see this by completing squares. Substitution gives

$$A\left[\left(\frac{x'}{1-z'}\right)^2 + \left(\frac{y'}{1-z'}\right)^2\right] + B\frac{x'}{1-z'} + C\frac{y'}{1-z'} + D = 0,$$

which simplifies to

$$A(x'^2 + y'^2) + Bx'(1-z') + Cy'(1-z') + D(1-z')^2 = 0.$$

Since $x'^2 + y'^2 + z'^2 = 1$, we have

$$A(1 - z'^2) + Bx'(1-z') + Cy'(1-z') + D(1-z')^2 = 0,$$

which upon division by $1 - z'$ yields

$$A(1 + z') + Bx' + Cy' + D(1-z') = 0,$$

the equation of a plane. This is obvious if the lircle is a line. The intersection of a plane and the sphere is a circle. Finally, the condition that the plane intersects the circle is the nondegeneracy condition for lircles. To see this, use the formula for distance from a point to a plane.

It is clear that the steps are reversible and hence all circles on the sphere are stereographic images of lircles in the complex plane.

A nice property of stereographic projection is that it preserves angles. Suppose that $ax + by + c = 0$ and $dx + ey + f = 0$ are two lines in the complex plane. We know that their images under stereographic projection are circles that intersect at the North Pole and another point. The angles of the curves at the intersections are the same (by symmetry), so let us determine the angle of intersection at the North Pole. From the equation for the plane of stereographic projection of a lircle, we see that the stereographic projections of the lines lie on the planes $ax' + by' + c(1 - z') = 0$ and $dx' + ey' + f(1 - z') = 0$, respectively. A tangent to a circle lying on a sphere lies in the tangent plane to the sphere at the point of tangency. Hence, tangents to the circles lie in the plane $z = 1$; they are given by $ax' + by' = 0, z = 1$ and $dx' + ey' = 0, z = 1$. It is evident from the form of the equations that the angle between the tangent lines is the same as the angle between the original lines.

The Riemann sphere has many pleasing properties. Any two lines intersect at ∞. A neighborhood of ∞ is a spherical cap on the Riemann sphere; it corresponds, under stereographic projection, to the outside of a circle centered at the origin. Thus, any path that moves away from the origin (a ray or spiral, for example) is said to tend to ∞. This is different from the real line, where we distinguishes between $+\infty$ and $-\infty$. Another nice property of the Riemann sphere is that stereographic projections of the points z and $1/\bar{z}$ are antipodal.

The symmetries (self-similarities) of the Riemann sphere are the *Möbius functions*, also called *linear fractional transformations*, of the form

$$z \mapsto \frac{az + b}{cz + d}, \quad a, b, c, d \in \mathbf{C}.$$

3

Captivating Formulas

> Mathematicians do not study objects, but relations among objects; they
> are indifferent to the replacement of objects by others as long as the re-
> lations don't change. Matter is not important, only form interests them.
> —HENRI POINCARÉ (1854–1912)

Mathematical formulas, whether simple or complicated, convey in symbols the essence
of mathematicians' discoveries. Some formulas are well known, such as Euler's formula
$e^{i\theta} = \cos\theta + i\sin\theta$. Some are less known. We will look at a few formulas I find beautiful,
some stark and some ornate. You may find them beautiful too.

3.1 Arithmetical Wonders

Here are three arithmetical curiosities:

$$123456789 \times 8 + 9 = 987654321$$

$$123456789 \times 9 + 10 = 1111111111$$

$$111111111 \times 111111111 = 12345678987654321.$$

It is easy to verify their truth, but why do they work? What happens when you do the
multiplication?

3.2 Heron's Formula and Heronian Triangles

Heron's formula, discovered by Heron of Alexandria (c. 10–70), gives the area of a triangle in terms of its side lengths. Suppose that a triangle has side lengths a, b, c, and
semiperimeter $s = (a + b + c)/2$. Then the area of the triangle is

$$A = \sqrt{s(s-a)(s-b)(s-c)}.$$

For instance, a triangle with sides 10, 11, and 13 has semiperimeter $s = 17$ and area

$$\sqrt{17 \cdot 7 \cdot 6 \cdot 4} = 2\sqrt{714}.$$

We will prove Heron's formula. Let vectors **a** and **b** represent the sides of lengths a and b. By the determinant formula of Chapter 1, the area of the triangle is

$$A = \frac{1}{2}|\det M|,$$

where M is the 2×2 matrix whose rows are **a** and **b**. Since the transpose M^t of M has the same determinant,

$$4A^2 = \det M \det M^t = \det(MM^t) = \begin{vmatrix} a^2 & \mathbf{a} \cdot \mathbf{b} \\ \mathbf{a} \cdot \mathbf{b} & b^2 \end{vmatrix}.$$

The third side of the triangle is represented by the vector $\mathbf{c} = \mathbf{a} - \mathbf{b}$, and

$$c^2 = (\mathbf{a} - \mathbf{b}) \cdot (\mathbf{a} - \mathbf{b}) = a^2 - 2\mathbf{a} \cdot \mathbf{b} + b^2.$$

Solving for $\mathbf{a} \cdot \mathbf{b}$, we obtain

$$4A^2 = \begin{vmatrix} a^2 & (a^2 + b^2 - c^2)/2 \\ (a^2 + b^2 - c^2)/2 & b^2 \end{vmatrix},$$

$$16A^2 = \begin{vmatrix} 2a^2 & a^2 + b^2 - c^2 \\ a^2 + b^2 - c^2 & 2b^2 \end{vmatrix}$$

$$= 4a^2b^2 - (a^2 + b^2 - c^2)^2$$

$$= (2ab + a^2 + b^2 - c^2)(2ab - a^2 - b^2 + c^2)$$

$$= ((a + b)^2 - c^2)(c^2 - (a - b)^2)$$

$$= (a + b + c)(a + b - c)(c + a - b)(c - a + b),$$

$$A^2 = s(s - c)(s - b)(s - a),$$

$$A = \sqrt{s(s - a)(s - b)(s - c)}.$$

A *Heronian triangle* is a triangle with rational side lengths and area. An example is the familiar right triangle with sides 3, 4, 5, and area 6. We will give a formula that generates all Heronian triangles.

Since the area of a triangle is half of its base times its height, the altitudes of a Heronian triangle are rational. Hence, we may scale a Heronian triangle by a rational factor so that it has an altitude of 2. We will assume that this altitude is to a longest side of the triangle. Thus, the triangle splits into two right triangles, as in the diagram.

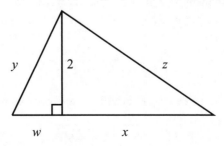

By hypothesis, the lengths $w + x$, y, and z are rational. We will show that w and x are rational. By the Pythagorean theorem,

$$w^2 + 4 = y^2$$
$$x^2 + 4 = z^2.$$

Subtraction gives

$$(w + x)(w - x) = y^2 - z^2,$$

and hence $w - x$ is rational. It follows that $(w - x) + (w + x) = 2w$ is rational, and thus w and x are rational.

From $w^2 + 4 = y^2$, we have $4 = (y + w)(y - w)$. Set

$$y + w = 2p$$

$$y - w = \frac{2}{p},$$

where p is a rational number. By the triangle inequality, $y + w > 2$ and so $p > 1$. Solving for y and w, we obtain

$$w = p - \frac{1}{p}, \; y = p + \frac{1}{p}, \quad p > 1.$$

This gives rational side lengths for the right triangle on the left in the diagram. The other right triangle has a similar form:

$$x = q - \frac{1}{q}, \; z = q + \frac{1}{q}, \quad q > 1.$$

Allowing for a rational scaling factor of r, every Heronian triangle has side lengths given uniquely, up to an interchange of p and q, by

$$r\left(p + \frac{1}{p}\right), \; r\left(q + \frac{1}{q}\right), \; r\left(p - \frac{1}{p} + q - \frac{1}{q}\right), \quad p, q > 1, r > 0,$$

with area

$$r^2\left(p - \frac{1}{p} + q - \frac{1}{q}\right).$$

For instance, the Heronian triangle corresponding to $p = 7/2, q = 13/5$, and $r = 10/19$ has side lengths

$$\frac{265}{133}, \frac{388}{287}, \frac{4941}{1729},$$

and area

$$\frac{49410}{32851}.$$

The 3–4–5 right triangle comes from $p = 2, q = 3, r = 6/5$.

Many questions can be asked about Heronian triangles. For example, can we find all Heronian triangles with consecutive integer side lengths? (easy) Are there Heronian triangles whose medians are rational numbers? (unsolved) A gem about Heronian triangles is that they can be scaled so that their vertices have integer coordinates in the plane (see [55]).

3.3   Sine, Cosine, and Exponential Function Expansions

The power series expansions of the sine, cosine, and exponential functions have esthetic appeal. A *power series expansion* of a function is an infinite series of the form

$$a_0 + a_1x + a_2x^2 + a_3x^3 + a_4x^4 + \cdots,$$

where the a_n are numbers and x is a variable.

How do we find the power series expansion of e^x? Assume that

$$e^x = a_0 + a_1x + a_2x^2 + a_3x^3 + a_4x^4 + \cdots,$$

for all real numbers x. If we let $x = 0$, then the right side of this expression collapses to a_0, while the left side is $e^0 = 1$. Hence $a_0 = 1$. Now we know that

$$e^x = 1 + a_1x + a_2x^2 + a_3x^3 + a_4x^4 + \cdots.$$

To determine a_1, differentiate both sides. As the derivative of e^x is e^x, we obtain

$$e^x = a_1 + 2a_2x + 3a_3x^2 + 4a_4x^3 + \cdots.$$

Letting $x = 0$, we have $1 = a_1$, so

$$e^x = 1 + 2a_2x + 3a_3x^2 + 4a_4x^3 + 5a_5x^4 + \cdots.$$

Taking another derivative, we obtain

$$e^x = 2a_2 + 3 \cdot 2a_3x + 4 \cdot 3a_4x^2 + 5 \cdot 4a_5x^3 + \cdots.$$

Letting $x = 0$, we have $1 = 2a_2$, so $a_2 = 1/2$. Repeating, we obtain a power series expansion for the exponential function:

$$e^x = 1 + x + \frac{x^2}{2!} + \frac{x^3}{3!} + \frac{x^4}{4!} + \cdots.$$

The series converges for all real numbers x. In fact, the variable can be any complex number z. Setting $x = 1$, we obtain a formula for e, the base of natural logarithms:

$$e = 1 + 1 + \frac{1}{2!} + \frac{1}{3!} + \frac{1}{4!} + \cdots \doteq 2.71828.$$

If we do the same for sine and cosine, we find that

$$\sin x = x - \frac{x^3}{3!} + \frac{x^5}{5!} - \frac{x^7}{7!} + \cdots$$

$$\cos x = 1 - \frac{x^2}{2!} + \frac{x^4}{4!} - \frac{x^6}{6!} + \cdots.$$

Leonhard Euler (1707–1783) observed that the expansion for the exponential function works just as well if x is a complex variable, and if we replace x by $i\theta$, where i is the imaginary unit ($i^2 = -1$), then we have a relation among the exponential, sine, and cosine functions:

$$e^{i\theta} = \cos\theta + i\sin\theta.$$

Letting $\theta = \pi$ yields

$$e^{i\pi} = \cos \pi + i \sin \pi = -1,$$

and hence

$$e^{i\pi} + 1 = 0.$$

This relation unites five important mathematical constants, $\pi, e, i, 1$, and 0, in one formula.

3.4 Tangent and Secant Function Expansions

We found in the previous section that the power series expansions of $\sin x$ and $\cos x$ follow a simple pattern. What about the power series expansions of $\tan x$ and $\sec x$? They are

$$\tan x = x + \frac{2x^3}{3!} + \frac{16x^5}{5!} + \frac{272x^7}{7!} + \frac{7936x^9}{9!} + \cdots$$

$$\sec x = 1 + \frac{x^2}{2!} + \frac{5x^4}{4!} + \frac{61x^6}{6!} + \frac{1385x^8}{8!} + \cdots.$$

What is the pattern of the sequences $\{1, 2, 16, 272, 7936, \ldots\}$ and $\{1, 1, 5, 61, 1385, \ldots\}$?

Suppose that

$$\tan x = \frac{a_0 x^0}{0!} + \frac{a_1 x^1}{1!} + \frac{a_2 x^2}{2!} + \frac{a_3 x^3}{3!} + \cdots$$

$$\sec x = \frac{b_0 x^0}{0!} + \frac{b_1 x^1}{1!} + \frac{b_2 x^2}{2!} + \frac{b_3 x^3}{3!} + \cdots.$$

From the differentiation formula $(\tan x)' = \sec^2 x$, we obtain

$$a_1 + \frac{a_2}{1!}x + \frac{a_3}{2!}x^2 + \cdots$$

$$= b_0^2 + \left(\frac{b_0}{0!} \frac{b_1}{1!} + \frac{b_1}{1!} \frac{b_0}{0!} \right) x + \left(\frac{b_0}{0!} \frac{b_2}{2!} + \frac{b_1}{1!} \frac{b_1}{1!} + \frac{b_2}{2!} \frac{b_0}{0!} \right) x^2 + \cdots.$$

Equating coefficients of like powers of x, we have

$$\frac{a_n}{(n-1)!} = \sum_{k=0}^{n-1} \frac{b_k}{k!} \frac{b_{n-1-k}}{(n-1-k)!}, \quad n \geq 1,$$

and hence

$$a_n = \sum_{k=0}^{n-1} \binom{n-1}{k} b_k b_{n-1-k}, \quad n \geq 1.$$

Similarly, from $(\sec x)' = \sec x \tan x$, we get

$$b_n = \sum_{k=0}^{n-1} \binom{n-1}{k} a_k b_{n-1-k}, \quad n \geq 1.$$

These recurrence relations, with the initial values $a_0 = 0$ and $b_0 = 1$, generate the sequences $\{a_n\}$ and $\{b_n\}$. You can show that $a_n = 0$ for n even, and $b_n = 0$ for n odd (which is expected since tan is an odd function and sec is an even function). Ignoring the 0s, we have the sequences $\{1, 2, 16, 272, 7936, \ldots\}$ and $\{1, 1, 5, 61, 1385, \ldots\}$.

We might ask whether the sequences have any other significance, and they do. They give the number of alternating permutations of the set $\{1, 2, \ldots, n\}$. An *alternating permutation* is a permutation in which the elements alternately increase and decrease. Let c_n be the number of alternating permutations of $\{1, 2, \ldots, n\}$. The following table lists the alternating permutations for $1 \leq n \leq 5$ and the corresponding values of c_n.

n	alternating permutations	c_n
1	1	1
2	12	1
3	132, 231	2
4	1324, 1423, 2314, 2413, 3412	5
5	13254, 14253, 14352, 15243, 15342, 23154, 24153, 24351,	16
	25143, 25341, 34152, 34251, 35142, 35241, 45132, 45231	

We can guess from these numbers that

$$c_n = \begin{cases} a_n & \text{if } n \text{ is odd} \\ b_n & \text{if } n \text{ is even.} \end{cases}$$

Can you prove this formula by mathematical induction?

3.5 Series for Pi

The famous number π, also called *Archimedes' constant*, is defined as the ratio of a circle's circumference to its diameter. The "p" in π ("pi") stands for "peripheria," the circumference of the circle.

The series formula[1]

$$\frac{\pi}{4} = \frac{1}{1} - \frac{1}{3} + \frac{1}{5} - \frac{1}{7} + \frac{1}{9} - \frac{1}{11} + \cdots.$$

is a striking representation of π.

We can derive it by starting with the geometric series formula

$$\frac{1}{1 + x^2} = 1 - x^2 + x^4 - \cdots + (-1)^n x^{2n} + (-1)^{n+1} \frac{x^{2n+2}}{1 + x^2}.$$

Thw expansion is valid for all real numbers x and all integers $n \geq 0$. Integrate both sides:

$$\int_0^1 \frac{dx}{1 + x^2} = \int_0^1 dx - \int_0^1 x^2 \, dx + \int_0^1 x^4 \, dx - \cdots + (-1)^n \int_0^1 x^{2n} \, dx$$

$$+ (-1)^{n+1} \int_0^1 \frac{x^{2n+2} \, dx}{1 + x^2}.$$

The left side evaluates to

$$\tan^{-1} 1 - \tan^{-1} 0 = \frac{\pi}{4} - 0 = \frac{\pi}{4},$$

---

[1]This formula was first discovered by Madhavan of Sangamagramam (1350–1425), and rediscovered by Gottfried Leibniz (1646–1716) and James Gregory (1638–1675).

and the right side to

$$1 - \frac{1}{3} + \frac{1}{5} - \cdots + (-1)^n \frac{1}{2n+1}$$

plus or minus the last integral, which is bounded:

$$0 < \int_0^1 \frac{x^{2n+2} \, dx}{1 + x^2} < \int_0^1 x^{2n+2} \, dx = \frac{1}{2n+3}.$$

Since the upper bound tends to 0 as $n \to \infty$, the integral tends to 0, and this finishes the derivation.

The series converges slowly. It requires 625 terms to obtain the approximation 3.14 for π. The difference between π and a partial sum of the series has the asymptotic formula

$$\pi - 4 \sum_{k=1}^n \frac{(-1)^{k+1}}{2k-1} \sim \sum_{m=0}^\infty \frac{b_{2m}}{(2n)^{2m}},$$

where $\{b_n\}$ is the sequence of numbers associated with the coefficients of the secant function, given in Tangent and Secant Function Expansions.

3.6   Product for Pi

Wallis's formula[2] for π is an infinite product:

$$\frac{\pi}{2} = \frac{2 \cdot 2}{1 \cdot 3} \cdot \frac{4 \cdot 4}{3 \cdot 5} \cdot \frac{6 \cdot 6}{5 \cdot 7} \cdots.$$

We can derive it from an infinite product expansion of the sine function:

$$\sin x = x \left(1 - \frac{x}{\pi}\right)\left(1 + \frac{x}{\pi}\right)\left(1 - \frac{x}{2\pi}\right)\left(1 + \frac{x}{2\pi}\right)\left(1 - \frac{x}{3\pi}\right)\left(1 + \frac{x}{3\pi}\right) \cdots.$$

Although we don't give a rigorous proof of this infinite product expansion, you can intuitively see that it's true because the zeros of $\sin x$ are $\pm n\pi$.

Letting $x = \pi/2$, we obtain

$$1 = \frac{\pi}{2}\left(1 - \frac{1}{2}\right)\left(1 + \frac{1}{2}\right)\left(1 - \frac{1}{4}\right)\left(1 + \frac{1}{4}\right)\left(1 - \frac{1}{6}\right)\left(1 + \frac{1}{6}\right) \cdots.$$

Therefore

$$\frac{\pi}{2} = \frac{2 \cdot 2}{1 \cdot 3} \cdot \frac{4 \cdot 4}{3 \cdot 5} \cdot \frac{6 \cdot 6}{5 \cdot 7} \cdots.$$

See [9] for more rigorous approaches.

Comparing the product and the series formulas for $\sin x$, we obtain (as Leonhard Euler did) a value of a series. We have

$$x - \frac{x^3}{3!} + \frac{x^5}{5!} - \frac{x^7}{7!} + \cdots$$

$$= x \left(1 - \frac{x}{\pi}\right)\left(1 + \frac{x}{\pi}\right)\left(1 - \frac{x}{2\pi}\right)\left(1 + \frac{x}{2\pi}\right)\left(1 - \frac{x}{3\pi}\right)\left(1 + \frac{x}{3\pi}\right) \cdots$$

$$= x \left(1 - \frac{x^2}{\pi^2}\right)\left(1 - \frac{x^2}{2^2\pi^2}\right)\left(1 - \frac{x^2}{3^2\pi}\right) \cdots.$$

---

[2]This formula was discovered by John Wallis (1616–1703), who made numerous contributions in algebra, geometry and calculus.

Equating coefficients of x^3 yields

$$-\frac{1}{6} = -\frac{1}{\pi^2} - \frac{1}{2^2\pi^2} - \frac{1}{3^2\pi^2} - \cdots,$$

and so

$$\frac{\pi^2}{6} = \frac{1}{1^2} + \frac{1}{2^2} + \frac{1}{3^2} + \cdots.$$

We will see how to compute further values of the sum

$$\sum_{m=1}^{\infty} \frac{1}{m^k}, \quad k \geq 2,$$

in The Zeta Function and Bernoulli Numbers.

3.7 Fibonacci Numbers and Pi

The *Fibonacci sequence*[3] $\{F_n\}$ is defined by

$$F_0 = 0, \ F_1 = 1, \quad F_n = F_{n-1} + F_{n-2}, \ n \geq 2.$$

The Fibonacci sequence is

$$0, \ 1, \ 1, \ 2, \ 3, \ 5, \ 8, \ 13, \ 21, \ 34, \ 55, \ \ldots.$$

It may be surprising that

$$\frac{\pi}{4} = \sum_{n=1}^{\infty} \tan^{-1} \frac{1}{F_{2n+1}}.$$

What do the Fibonacci numbers have to do with π? The partial sums are

$$\sum_{n=1}^{k} \tan^{-1} \frac{1}{F_{2n+1}} = \frac{\pi}{4} - \tan^{-1} \frac{1}{F_{2k+2}},$$

and the summation result follows instantly. The partial sums formula can be derived using the formula for the tangent of a difference and Cassini's identity

$$F_n^2 - F_{n+1}F_{n-1} = (-1)^{n+1}, \quad n \geq 1.$$

See [28] for a wealth of identities involving Fibonacci numbers.

3.8 Volume of a Ball

A circle of radius r has area πr^2. A sphere of radius r has volume $\frac{4}{3}\pi r^3$. Does it make sense to talk about the volume of a higher-dimensional sphere?

A d-dimensional *ball* of radius r, centered at the origin, is the set of points $(x_1, \ldots, x_d)$ in d-dimensional Euclidean space such that

$$x_1^2 + \cdots + x_d^2 \leq r^2.$$

---

[3]The Fibonacci sequence was introduced by Leonardo of Pisa, known as Fibonacci (c. 1170–c. 1250), one of the most innovative mathematicians of the Middle Ages.

The boundary of a ball is a hypersphere. In dimension 1 we have a line segment, in dimension 2 a disk (a circle and its interior), and in dimension 3 an ordinary ball. A 4-dimensional ball is more difficult to picture, but we can use analogies with the lower-dimensional objects.

We will show that the volume of a d-dimensional ball of radius r, for $d \geq 1$, is

$$V_d \, r^d,$$

where V_d, the volume of the unit d-dimensional ball (radius 1), is

$$\frac{\pi^{d/2}}{(d/2)!}.$$

Though we cannot construct or even visualize an n-dimensional sphere, we can find its volume!

If d is even, then $d/2$ is an integer and $(d/2)!$ is the usual factorial function. If d is odd, we have to explain what $(d/2)!$ means. We define

$$\left(\frac{1}{2}\right)! = \frac{\sqrt{\pi}}{2}.$$

An explanation for this, using the gamma function (a generalization of the factorial function), will be given in the next section. Now we can compute $(d/2)!$ for d odd by treating the expression like a normal factorial, multiplying $d/2$ by $d/2 - 1$, $d/2 - 2$, etc., until we get down to $1/2$. For instance,

$$\left(\frac{7}{2}\right)! = \frac{7}{2} \cdot \frac{5}{2} \cdot \frac{3}{2} \cdot \left(\frac{1}{2}\right)! = \frac{105\sqrt{\pi}}{16}.$$

We already know that $V_1 = 2$, $V_2 = \pi$, and $V_3 = (4/3)\pi$. We will check V_2 and V_3, and then determine V_4.

Here is a check of the formula for the area of a unit circle. We use polar coordinates:

$$V_2 = \int_0^{2\pi}\int_0^1 r \, dr \, d\theta = \int_0^{2\pi} \frac{1}{2}r^2 \Big|_0^1 \, d\theta = \frac{1}{2}\int_0^{2\pi} d\theta = \frac{1}{2}(2\pi) = \pi.$$

For the unit ball in 3-dimensional space, we use polar coordinates to represent the first two dimensions. We have $x_1^2 + x_2^2 = r^2$, and x_3 is bounded by a line segment of radius $\sqrt{1 - r^2}$. Summing (with an integral) the volumes of the line segments,

$$V_3 = \int_0^{2\pi}\int_0^1 2\sqrt{1 - r^2}\, r \, dr \, d\theta = -\frac{2}{3}(1 - r^2)^{3/2}\Big|_0^1 (2\pi) = \frac{4}{3}\pi.$$

For the 4-dimensional ball,

$$V_4 = \int_0^{2\pi}\int_0^1 \pi\left(\sqrt{1 - r^2}\right)^2 r \, dr \, d\theta = \pi\left(\frac{1}{2}r^2 - \frac{1}{4}r^4\right)\Big|_0^1 (2\pi) = \frac{1}{2}\pi^2.$$

Thus $V_4 = \pi^2/2$.

Let's prove the volume formula for the d-dimensional unit ball. Representing the first two dimensions by polar coordinates r and θ, the cross-section is a $(d-2)$-dimensional ball of radius $\sqrt{1-r^2}$. Hence

$$V_d = \int_0^{2\pi} \int_0^1 V_{d-2} \left(\sqrt{1-r^2} \right)^{d-2} r \, dr \, d\theta.$$

(Using the letter 'd' to denote both a differential and a dimension should cause no confusion.) The double integral is

$$V_{d-2} \left(-\frac{1}{d} \right) (1-r^2)^{d/2} \Big|_0^1 (2\pi) = V_{d-2} \left(\frac{2\pi}{d} \right).$$

Therefore we can find the constants V_d recursively:

$$V_d = V_{d-2} \left(\frac{2\pi}{d} \right), \quad d \geq 2.$$

The volume formula follows at once by mathematical induction.

We will use it to prove Lagrange's four squares theorem in Chapter 4.

3.9   Euler's Integral Formula

In Volume of a Ball, we said that to evaluate the factorial of a half-integer we need the *gamma function*. Leonhard Euler (1707–1783) noted that

$$n! = \int_0^\infty t^n e^{-t} \, dt, \quad n \geq 0.$$

This can be obtained by starting with

$$\int_0^\infty e^{-at} \, dt = \left(-\frac{e^{-at}}{a} \right) \Big|_{t=0}^\infty = \frac{1}{a},$$

and differentiating n times with respect to a to get

$$\int_0^\infty (-1)^n t^n e^{-at} \, dt = (-1)^n \frac{n!}{a^{n+1}}.$$

Setting $a = 1$ yields Euler's result. The gamma function is defined for all complex numbers z with positive real part by

$$\Gamma(z) = \int_0^\infty t^{z-1} e^{-t} \, dt, \quad \Re z > 0.$$

For integers,

$$\Gamma(n) = (n-1)!, \quad n \geq 1.$$

We will show that $(1/2)! = \sqrt{\pi}/2$. From the definition of $\Gamma(z)$, we have

$$\left(\frac{1}{2} \right)! = \Gamma \left(\frac{3}{2} \right) = \int_0^\infty t^{1/2} e^{-t} \, dt.$$

The change of variables $t = x^2$ gives

$$\left(\frac{1}{2}\right)! = 2 \int_0^\infty x^2 e^{-x^2}\, dx,$$

and integration by parts (with $u = x$ and $dv = xe^{-x^2}\, dx$) gives

$$\left(\frac{1}{2}\right)! = \left(-xe^{-x^2}\right)\Big|_0^\infty + \int_0^\infty e^{-x^2}\, dx.$$

The first term on the right is 0, and we use the integral given in A Classic Integral in Chapter 5:

$$\int_{-\infty}^\infty e^{-x^2}\, dx = \sqrt{\pi}.$$

Since e^{-x^2} is an even function, $(1/2)!$ is half of this integral.

3.10   Euler's Polyhedral Formula

A cube has eight vertices, twelve edges, and six faces.

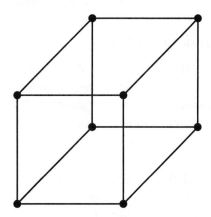

For a convex polyhedron, let V be the number of vertices, E the number of edges, and F the number of faces. Euler's polyhedral formula is

$$V - E + F = 2.$$

You could test this formula on some other examples, say a tetrahedron and an octahedron.

Consider the polyhedron below. It is a cube with a square tunnel from one face to the opposite face.

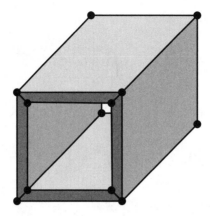

The polyhedron is drawn as a solid, so we can't see all its vertices, edges, and faces. But we can infer that $V = 16$, $E = 32$ and $F = 16$. So

$$V - E + F = 0.$$

The hole makes the difference. If a polyhedron can be drawn on a surface with g holes (g is called the *genus* of the surface), then Euler's formula is

$$V - E + F = 2 - 2g.$$

Care must be taken in the definition of genus. It applies only to surfaces that are orientable. A Möbius strip, a surface with only one side, is not orientable. For a proof of Euler's polyhedral formula, see [1]. For generalizations of Euler's formula, a rich resource is [32].

3.11   The Smallest Taxicab Number

A story is often told about G. H. Hardy (1887–1947) visiting Srinivasa Ramanujan (1887–1920) when Ramanujan was ill in a hospital. Hardy said that the number of his taxicab, 1729, seemed to him to be a dull number. Ramanujan responded that, on the contrary, 1729 is the smallest number that is the sum of two cubes in two ways. That is,

$$1729 = 10^3 + 9^3 = 12^3 + 1^3.$$

However, the statement that 1729 is the smallest such number must be altered if we allow cubes of negative numbers. Let's find the smallest positive integer that is the sum of two cubes (positive or negative) in two ways. We list the first ten cubes.

n	1	2	3	4	5	6	7	8	9	10
n^3	1	8	27	64	125	216	343	512	729	1000

From the curious relation

$$3^3 + 4^3 + 5^3 = 6^3,$$

we have

$$91 = 3^3 + 4^3 = 6^3 + (-5)^3,$$

so 91 is the sum of two cubes in two ways. This is the smallest such positive integer, as can be verified using the table.

3.12   Infinity and Infinity Squared

The collection of all *whole numbers* $\{0, 1, 2, 3, 4, 5, \ldots\}$ is an infinite set. The collection of all ordered pairs of whole numbers,

$$\{(0, 0), (0, 1), (0, 2), \ldots, (1, 0), (1, 1), (1, 2), \ldots\},$$

is also infinite. Does it makes sense to talk about the size of an infinite set, to say that two infinite sets are the same size or that one is larger than the other? Georg Cantor (1845–1918) described a way to do so. The key definition is that two sets, finite or infinite, are the same size if there is a bijection (a one-to-one correspondence) between them. Although it may appear that the set of ordered pairs of whole numbers is larger than the set of whole numbers, the two infinite sets are the same size.

The mapping

$$(m, n) \mapsto \frac{(m + n)(m + n + 1)}{2} + m, \quad m, n \geq 0$$

takes an ordered pair of whole numbers to a whole number. The correspondence is shown in the table.

	n							
	0	1	2	3	4	5	6	...
m 0	0	1	3	6	10	15	21	...
1	2	4	7	11	16	22	29	...
2	5	8	12	17	23	30	38	...
3	9	13	18	24	31	39	48	...
4	14	19	25	32	40	49	59	...
5	20	26	33	41	50	60	71	...
6	27	34	42	51	61	72	84	...
⋮	⋮	⋮	⋮	⋮	⋮	⋮	⋮	

The table contains all the whole numbers, increasing in order along diagonals. Given a whole number w, let n be the unique whole number such that

$$\frac{n(n + 1)}{2} \leq w < \frac{(n + 1)(n + 2)}{2}.$$

Then let m be the whole number defined by

$$m = w - \frac{n(n + 1)}{2}.$$

The mapping $w \mapsto (m, n)$ is the other half of the bijection. Thus, the set of all whole numbers has just as many elements as the set of all ordered pairs of whole numbers.

3.13   Complex Functions

Consider the function

$$f(z) = z^2,$$

from $\mathbf{C}$ to $\mathbf{C}$. Each complex number z is squared to produce a new complex number z^2. We can't graph the function in two dimension because there are four dimensions in play, two for the domain and two for the range. However, we can say what the squaring function does geometrically.

We can represent a complex number z as

$$z = re^{i\theta},$$

where $r = |z|$ and θ is the angle that z makes in the complex plane, measured counterclockwise from the positive real axis. Because

$$z^2 = r^2 e^{2i\theta}$$

is a vector with length r^2 and argument 2θ, the squaring map changes the length of vectors and rotates them counterclockwise.

The *derivative* of a complex function is defined in the same way as for a real-valued function:

$$f'(z) = \lim_{\Delta z \to 0} \frac{f(z + \Delta z) - f(z)}{\Delta z}.$$

But there is a subtle difference that turns out to be critical. When we say that $\Delta z \to 0$, the complex number Δz can approach 0 in any fashion, and this may prevent the existence of the derivative. This restriction on the existence of the derivative means that complex functions that have a derivative are special.

It is possible for a real-valued function to have a derivative that does not have a derivative. Let

$$f(x) = x^{5/3}.$$

Then $f'(x) = (5/3)x^{2/3}$, but $f'(x)$ is not differentiable at $x = 0$. However, if a function of a complex variable is differentiable, then all its higher derivatives exist.

If $f(z) = z^2$, then

$$\begin{aligned}
f'(z) &= \lim_{\Delta z \to 0} \frac{(z + \Delta z)^2 - z^2}{\Delta z} \\[2mm]
&= \lim_{\Delta z \to 0} \frac{z^2 + 2z\Delta z + (\Delta z)^2 - z^2}{\Delta z} \\[2mm]
&= \lim_{\Delta z \to 0} \frac{2z\Delta z + (\Delta z)^2}{\Delta z} \\[2mm]
&= \lim_{\Delta z \to 0} (2z + \Delta z) \\[2mm]
&= 2z.
\end{aligned}$$

We found that $(z^2)' = 2z$, as for real-valued functions.

Consider the complex conjugation function, $f(z) = \bar{z}$. Its derivative would be

$$f'(z) = \lim_{\Delta z \to 0} \frac{\overline{(z + \Delta z)} - \bar{z}}{\Delta z}.$$

Writing $z = x + iy$ and $\Delta z = \Delta x + i\,\Delta y$, we have

$$f'(z) = \lim_{\Delta x, \Delta y \to 0} \frac{\overline{(x + iy + \Delta x + i\,\Delta y)} - \overline{(x + iy)}}{\Delta x + i\,\Delta y}$$

$$= \lim_{\Delta x, \Delta y \to 0} \frac{(x + \Delta x - iy - i\,\Delta y) - (x - iy)}{\Delta x + i\,\Delta y}$$

$$= \lim_{\Delta x, \Delta y \to 0} \frac{\Delta x - i\,\Delta y}{\Delta x + i\,\Delta y}.$$

If we set $\Delta y = 0$, then the limit would be 1. If we set $\Delta x = 0$, then the limit would be -1. But a limit must be unique, so it doesn't exist and therefore $f(z) = \bar{z}$ has no derivative.

Suppose that

$$f(z) = u + iv,$$

where u and v are functions of x and y, so $u = u(x, y)$ and $v = v(x, y)$. It can be shown that necessary and sufficient conditions for $f'(z)$ to exist are given by the *Cauchy–Riemann equations*

$$\frac{\partial u}{\partial x} = \frac{\partial v}{\partial y}, \quad \frac{\partial u}{\partial y} = -\frac{\partial v}{\partial x}.$$

These *partial derivatives* express the rate of change of the functions when only one of the variables changes. For example, $\partial u / \partial x$, the partial derivative of u with respect to x, is the derivative of $u(x, y)$ as a function of x, where y is held constant.

As an exercise, you can check that $f(z) = z^2$ satisfies the Cauchy–Riemann equations. The first step is to write $z = x + iy$.

We can represent a complex number $a + bi$ as a 2×2 matrix

$$\begin{bmatrix} a & -b \\ b & a \end{bmatrix},$$

a rotation and stretching matrix that mimics what happens when we multiply complex numbers by $a + bi$. Thus, we can think of multiplication by a complex number as a map from $\mathbf{C}$ to $\mathbf{C}$ or as a map from $\mathbf{R}^2$ to $\mathbf{R}^2$.

A map from $\mathbf{R}^2$ to $\mathbf{R}^2$ with

$$\begin{bmatrix} x \\ y \end{bmatrix} \mapsto \begin{bmatrix} u(x, y) \\ v(x, y) \end{bmatrix}$$

has a derivative that is at each point a linear map of the form

$$\begin{bmatrix} \dfrac{\partial u}{\partial x} & \dfrac{\partial u}{\partial y} \\[2mm] \dfrac{\partial v}{\partial x} & \dfrac{\partial v}{\partial y} \end{bmatrix}.$$

For a complex function, the Cauchy–Riemann equations ensure that the derivative map acts like multiplication by a complex number, for the corresponding matrix has the appropriate form:

$$\begin{bmatrix} \dfrac{\partial u}{\partial x} & -\dfrac{\partial v}{\partial x} \\[2mm] \dfrac{\partial v}{\partial x} & \dfrac{\partial u}{\partial x} \end{bmatrix}.$$

Differentiability is a stronger property for complex functions than for real-valued functions. If a complex function has a derivative in some region, then it is differentiable infinitely many times there. In the next section, we will see an example of a differentiable complex function defined on the entire complex plane except for one point.

3.14   The Zeta Function and Bernoulli Numbers

The *zeta function* is defined for all integers $k \geq 2$ by

$$\zeta(k) = \sum_{m=1}^{\infty} \frac{1}{m^k}.$$

(If $k = 1$ we have the divergent harmonic series.)

Bernoulli numbers[4] are defined recursively by $B_0 = 1$ and

$$B_n = -\frac{1}{n+1} \sum_{k=0}^{n-1} \binom{n+1}{k} B_k, \quad n \geq 1.$$

The first few Bernoulli numbers are

$$1, \ -\frac{1}{2}, \ \frac{1}{6}, \ 0, \ -\frac{1}{30}, \ 0, \ \frac{1}{42}, \ 0, \ -\frac{1}{30}, \ 0, \ \frac{5}{66}, \ \dots \ .$$

It appears that $B_{2n+1} = 0$, for $n \geq 1$, and this is true. Bernoulli numbers are related to the coefficients of the tangent function that we found in Tangent and Secant Function Expansions. See, e.g., [21, p. 287].

There is a connection between the zeta function and Bernoulli numbers. For any positive integer n, we have

$$\zeta(2n) = (-1)^{n+1} \frac{(2\pi)^{2n}}{2(2n)!} B_{2n}.$$

See, e.g., [23]. In Product for Pi, we discovered that

$$\sum_{m=1}^{\infty} \frac{1}{m^2} = \frac{\pi^2}{6}.$$

What is the value of

$$\sum_{m=1}^{\infty} \frac{1}{m^4} \ ?$$

---

[4]Bernoulli numbers are named for Jakob Bernoulli (1654–1705), who made important discoveries in probability theory and counting, and in calculus and differential equations.

3.15   The Riemann Zeta Function

We saw in The Zeta Function and Bernoulli Numbers that the zeta function is defined for all integers $k \geq 2$ by

$$\zeta(k) = \sum_{m=1}^{\infty} \frac{1}{m^k}.$$

We also know that

$$\zeta(2) = 1 + \frac{1}{2^2} + \frac{1}{3^2} + \frac{1}{4^2} + \cdots = \frac{\pi^2}{6}$$

and

$$\zeta(2n) = (-1)^{n+1} \frac{(2\pi)^{2n}}{2(2n)!} B_{2n}.$$

In 1859 Bernhard Riemann showed how to extend the definition of ζ to the entire complex plane except the point 1. For a complex number s with real part greater than 1, define

$$\zeta(s) = \sum_{m=1}^{\infty} \frac{1}{m^s}.$$

It can be shown that the sum converges.

The series

$$\sum_{m=1}^{\infty} \frac{(-1)^{m-1}}{m^s}.$$

converges when the real part of s is positive. Since

$$\zeta(s) - \sum_{m=1}^{\infty} \frac{(-1)^{m-1}}{m^s} = \sum_{m=1}^{\infty} \frac{1}{m^s} + \sum_{m=1}^{\infty} \frac{(-1)^m}{m^s}$$

$$= \sum_{n=1}^{\infty} \frac{2}{(2n)^s}$$

$$= 2^{1-s} \zeta(s),$$

we have

$$\zeta(s) = \frac{1}{1 - 2^{1-s}} \sum_{m=1}^{\infty} \frac{(-1)^{m-1}}{m^s}.$$

This extends the definition of $\zeta(s)$ to the half plane where the real part of s is positive. Riemann showed that the definition of ζ can be extended to the entire complex plane except $s = 1$ by using a functional equation

$$\zeta(s) = 2^s \pi^{s-1} \sin\left(\frac{\pi s}{2}\right) \Gamma(1-s) \zeta(1-s), \quad \Re s < 0.$$

Here $\Gamma(z)$ is the gamma function, a generalization of the factorial function that we met in Euler's Integral Formula:

$$\Gamma(z) = \int_0^{\infty} t^{z-1} e^{-t} \, dt.$$

Recall that $\Gamma(n) = (n-1)!$ when n is a positive integer.

As an exercise, you can use the functional equation to prove that

$$\zeta(-1) = -\frac{1}{12}.$$

The extended definition of $\zeta(s)$ is a differentiable function. According to the theory of complex functions, it is uniquely defined.

Riemann's zeta function is connected with prime numbers, as seen in the formula

$$\zeta(s) = \sum_{m=1}^{\infty} \frac{1}{m^s} = \prod_{p} \frac{1}{1 - p^s}, \quad \Re s > 1,$$

where the product ranges over all primes p. To understand why this formula holds, use the geometric series sum formula

$$\frac{1}{1 - p^s} = 1 + p^s + p^{2s} + p^{3s} + \cdots,$$

and observe what happens when such sums for various primes p are multiplied together.

3.16   The Jacobi Identity

Consider the set of $n \times n$ matrices with real entries. Although we can add and multiply matrices together, matrix multiplication is not commutative for $n > 1$. For example,

$$\begin{bmatrix} 1 & 1 \\ 1 & 2 \end{bmatrix} \begin{bmatrix} 1 & 2 \\ 1 & 1 \end{bmatrix} = \begin{bmatrix} 2 & 3 \\ 3 & 4 \end{bmatrix}$$

but

$$\begin{bmatrix} 1 & 2 \\ 1 & 1 \end{bmatrix} \begin{bmatrix} 1 & 1 \\ 1 & 2 \end{bmatrix} = \begin{bmatrix} 3 & 5 \\ 2 & 3 \end{bmatrix}.$$

However, we can define a multiplication of matrices that is *anti*-commutative. We define the product of two square matrices of the same size, A and B, in bracket notation as

$$[A, B] = AB - BA.$$

Then

$$[B, A] = BA - AB = -[A, B].$$

Under this definition of multiplication, the set of matrices satisfies the *Jacobi identity*[5]:

$$[A, [B, C]] + [B, [C, A]] + [C, [A, B]] = 0.$$

---

[5]The Jacobi identity is due to Carl Gustav Jacob Jacobi (1804–1851), one of the most influential mathematicians of the 1800s.

Let's check:

$$[A, [B, C]] + [B, [C, A]] + [C, [A, B]]$$

$$= [A, BC - CB] + [B, CA - AC] + [C, AB - BA]$$

$$= A(BC - CB) - (BC - CB)A + B(CA - AC)$$

$$- (CA - AC)B + C(AB - BA) - (AB - BA)C$$

$$= ABC - ACB - BCA + CBA + BCA - BAC$$

$$- CAB + ACB + CAB - CBA - ABC + BAC$$

$$= 0.$$

An *algebra* is a vector space in which we can multiply vectors. An algebra is *bilinear* if it satisfies

$$[ra + sb, c] = r[a, c] + s[b, c]$$

and

$$[a, rb + sc] = r[a, b] + s[a, c],$$

where r and s are scalars (elements of the base field) and a, b, and c are vectors. It is easy to show that the Jacobi product gives a bilinear algebra.

An algebra with a bilinear anticommutative multiplication that satisfies Jacobi's identity is called a *Lie algebra*, named after Sophus Lie (1842–1899). Lie algebras are nonassociative: the identity

$$[a, [b, c]] = [[a, b], c]$$

is not assumed.

Another, perhaps more familiar, example of a Lie algebra is the vector cross product, $[u, v] = u \times v$, defined on 3-dimensional Euclidean space. You can check that the vector cross product satisfies Jacobi's identity by direct calculation or by using the identity $a \times (b \times c) = (a \cdot c)b - (a \cdot b)c$, where a, b, and c are three-dimensional real vectors.

Lie algebras are important in quantum mechanics. A very understandable explanation of Lie algebras in the context of Euclidean space is [51].

3.17 Entropy

Suppose that you have a goose that lays golden eggs, silver eggs, and bronze eggs. She lays one egg each day and you don't know which kind it will be. Half of the days she lays golden eggs, one-fourth of the days she lays silver eggs, and one-fourth of the days she lays bronze eggs. Your neighbor has a goose who lays golden eggs, silver eggs, and bronze eggs with equal probability, one per day. There is uncertainty about what kind of eggs the geese will produce on any day. But how much uncertainty? Which goose is more unpredictable?

We will give a mathematical definition of uncertainty and use it to measure the uncertainty associated with your goose and your neighbor's goose.

We say that a *source S* is a set of outcomes that occur with various probabilities. Suppose that outcomes $x_1, x_2, \ldots, x_n$ occur with probabilities $p_1, p_2, \ldots, p_n$, respectively, where the probabilities are nonnegative real numbers that sum to 1.

In 1948 Claude E. Shannon (1916–2001), the founder of information theory, defined the *entropy H* of a source:

$$H(S) = -\sum_{i=1}^{n} p_i \log p_i.$$

Entropy is a weighted average of the logarithms of the probabilities of events. The logarithms normalize the probabilities so that the resulting calculations can be done in convenient units called *bits* of information. In information theory, calculations are done with base 2 logarithms.

Let's calculate the entropy of the two magical geese. Your goose lays eggs with entropy

$$H(\text{your goose}) = -\frac{1}{2}\log_2\frac{1}{2} - \frac{1}{4}\log_2\frac{1}{4} - \frac{1}{4}\log_2\frac{1}{4} = 1.5 \text{ bits.}$$

Your neighbor's goose lays eggs with entropy

$$H(\text{neighbor's goose}) = -\frac{1}{3}\log_2\frac{1}{3} - \frac{1}{3}\log_2\frac{1}{3} - \frac{1}{3}\log_2\frac{1}{3} \doteq 1.58 \text{ bits.}$$

Your neighbor's goose is more unpredictable than your goose by about 0.08 bits. Maximum entropy occurs when all outcomes are equally probable.

Shannon proved the two main theorems of information theory.

Shannon's first theorem says that given any information source, the most compact way to encode it with 0s and 1s requires, on average, a codeword whose length is equal to the entropy of the source. So, in our example of the geese that lay expensive eggs, if you want to keep a day-by-day journal account of the type of eggs that your goose lays, you need to expend, on average, 1.5 binary symbols per day, no matter what encoding scheme you devise. Your neighbor needs to use an average of 1.58 symbols per day.

Shannon's second theorem says that when information is sent over a noisy channel, where some symbols may be distorted, we can always devise a code to send the information at near perfect accuracy, but at a slower rate than if no code is used. The rate is given by a quantity called the channel capacity, which is defined in terms of entropy. See, e.g., [15].

Shannon didn't arrive at the information theory definition of entropy in a vacuum. Rudolf Julius Emanuel Clausius (1822–1888) introduced the concept of entropy in thermodynamics, Ludwig Eduard Boltzmann (1844–1906) gave a mathematical formulation, and Josiah Willard Gibbs (1839–1903) described entropy as an amount of randomness. The work of Shannon's predecessors aided him in formulating his ideas.

3.18   Rook Paths

A chess Rook can move any number of squares horizontally or vertically in one step. How many paths can a Rook take from the lower-left corner square to the upper-right corner square of an 8×8 chessboard, assuming that it moves right or up at each step? An example of a Rook path is shown in Figure 3.1.

We want to count lattice paths from $(0, 0)$ to (n, n) with steps of the form $(x, 0)$ or $(0, y)$, where x and y are positive integers.

The Rook path problem can be solved by generalizing to find the number of paths from $(0,0)$ to any square on an arbitrary size board, that is, to any point (m, n). Let $r(m, n)$ be the number of paths, where $m, n \geq 0$. We set $r(0,0) = 1$. By symmetry, $r(m, n) = r(n, m)$. For m or n positive, $r(m, n)$ is equal to the sum of the values of r for the horizontal and vertical predecessors of (m, n), since the Rook arrives at (m, n) from one of the squares to its left or below it. For example, $r(3, 2) = (2 + 5 + 14) + (4 + 12) = 37$. From the following table, we see that the number of Rook paths from the lower-left corner to the upper-right corner of an 8×8 chessboard is $r(7, 7) = 470010$.

⋮	⋮	⋮	⋮	⋮	⋮	⋮	⋮	
64	320	1328	4864	16428	52356	159645	470010	...
32	144	560	1944	6266	19149	56190	159645	...
16	64	232	760	2329	6802	19149	52356	...
8	28	94	289	838	2329	6266	16428	...
4	12	37	106	289	760	1944	4864	...
2	5	14	37	94	232	560	1328	...
1	2	5	12	28	64	144	320	...
1	1	2	4	8	16	32	64	...

We determined $r(m, n)$ using a variable number of preceding terms. But there is a recurrence relation that requires only three preceding terms:

$$r(0,0) = 1, \ r(0,1) = 1, \ r(1,0) = 1, \ r(1,1) = 2;$$

$$r(m, n) = 2r(m-1, n) + 2r(m, n-1) - 3r(m-1, n-1), \quad m \geq 2 \text{ or } n \geq 2.$$

(We assume that $r(m, n) = 0$ for m or n negative.) It can be proved by the method of inclusion and exclusion and is an exercise in Appendix B.

The recurrence formula yields a rational generating function for the doubly-infinite sequence $\{r(m, n)\}$, namely,

$$\sum_{m \geq 0, n \geq 0} r(m, n) x^m y^n = \frac{(1-x)(1-y)}{1 - 2(x + y) + 3xy}.$$

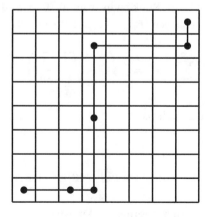

Figure 3.1. A Rook path.

The form of the denominator is given by the recurrence relation. The numerator is obtained by multiplying the denominator by the polynomial that represents the initial values, $1 + x + y + 2xy$, keeping only those monomials with exponents of x and y both less than 2.

Another way to obtain the generating function for Rook paths is to start with the generating function $1/(1 - x - y)$, which counts sequences of length n having some number of x's and a complementary number of y's (the total number of x's and y's is n). For Rook paths, we allow an arbitrary step length in either direction. This amounts to replacing x by $x/(1 - x)$ and y by $y/(1 - y)$. Hence, the generating function for Rook paths is

$$\frac{1}{1 - (x/(1 - x)) - (y/(1 - y))}.$$

We can generalize Rook paths to three dimensions. How many ways can a Rook move from $(0, 0, 0)$ to (m, n, o), where each step is a positive integer multiple of $(1, 0, 0)$, $(0, 1, 0)$, or $(0, 0, 1)$?

The generating function for three-dimensional Rook paths is

$$\frac{(1 - x)(1 - y)(1 - z)}{1 - 2(x + y + z) + 3(xy + yz + zx) - 4xyz}.$$

In dimension d, an asymptotic formula for the number of Rook paths from the origin to a main diagonal point is

$$r(n, \ldots, n) \sim (d + 1)^{dn-1} d^{(d+2)/2} (2\pi n(d + 2))^{(1-d)/2}.$$

See [16]. It is surprising that π appears in the formula.

Manuel Kauers and Doron Zeilberger have conjectured that, for n fixed, the number of Rook paths from the origin to a main diagonal point is

$$r(n, \ldots, n) \sim e^{n-1} \frac{(nd)!}{n!^d}.$$

Let $r_n = r(n, n)$, the number of Rook paths from $(0, 0)$ to the diagonal point (n, n). The generating function for the sequence $\{r_n\} = \{1, 2, 14, 106, 838, \ldots\}$ is

$$\sum_{n=0}^{\infty} r_n x^n = \frac{1}{2} \left(1 + \sqrt{\frac{1 - x}{1 - 9x}} \right).$$

A recurrence formula for such paths is

$$r_0 = 1, \; r_1 = 2;$$

$$r_n = ((10n - 6)r_{n-1} - (9n - 18)r_{n-2})/n, \quad n \geq 2.$$

No counting proof of this recurrence formula is known.

For three-dimensional Rook paths, the diagonal sequence $\{r_n = r(n, n, n)\}$ satisfies the recurrence formula

$$r_0 = 1, r_1 = 6, r_2 = 222, r_3 = 9918;$$

$$r_n = ((121n^3 - 212n^2 + 85n + 6)r_{n-1}$$
$$+ (475n^3 - 3462n^2 + 7853n - 5658)r_{n-2}$$
$$+ (-1746n^3 + 14580n^2 - 40662n + 37908)r_{n-3}$$
$$+ (1152n^3 - 12672n^2 + 46080n - 55296)r_{n-4})/(2n^3 - 2n^2), \quad n \geq 4.$$

Such recurrence relations exist for Rook paths to a diagonal point in any dimension, but their orders and the degrees of the polynomial coefficients are unknown in general.

A Rook path is equivalent to a game of Nim. In Nim, two players alternately remove any number of stones from one of a number of piles. The game ends when the last stone is removed. A Rook path from $(0, 0, \ldots, 0)$ to $(a_1, a_2, \ldots, a_d)$ is equivalent to a Nim game starting with d piles of stones of sizes $a_1, a_2, \ldots, a_d$.

4

Delightful Theorems

Mathematics is like looking at a house from different angles.

THOMAS F. STORER[1] (1938–2006)

Mathematicians prove theorems. Once a theorem is proved, it is true for all time. The theorems proved by the ancient Greeks are as true today as they were over two thousand years ago, and the theorems proved today will be true even if, after millions of years, humans evolve into another species. In this chapter we present some delightful and sometimes surprising theorems.

4.1 A Square inside Every Triangle

Given any triangle, is it always possible to inscribe a square in it? We require that the square has a side on one of the sides of the triangle, with the other two corners touching the other sides of the triangle.

The answer is yes, by similarity. Put the triangle on top of a square, as $\triangle ABC$ is placed in Figure 4.1. Now extend the other two sides of $\triangle ABC$ so that they meet the line that the square sits on. This results in a triangle similar to the given triangle and circumscribing the square. Finally, change the scale of the whole diagram so that the circumscribing triangle is the same size as our given triangle—and we are done. Note that the side of the triangle we place on the square must be chosen so that the altitude to that side lies inside the triangle. We have proved the theorem.

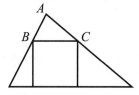

Figure 4.1. A square inscribed in a triangle.

Theorem. *Given any triangle, there exists a square inscribed inside it.*

---

[1] Tom Storer, the first Native American to earn a Ph.D. in mathematics, was the author's thesis advisor.

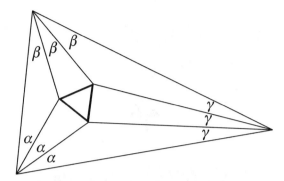

Figure 4.2. Morley's theorem (with the equilateral triangle in bold).

In the proof, we worked backwards, starting with the square to be inscribed and fitting a triangle around it. The idea of working backwards in a geometric construction is used strikingly in the proof of the next theorem.

By the way, the square inscribed in the triangle is constructible using straightedge and compass. How can you do the construction?

4.2 Morley's Theorem

One of the most delightful theorems of plane geometry was discovered fairly recently, a little over a hundred years ago.

Morley's Theorem. *In any triangle, the three points of intersection of adjacent angle trisectors are the vertices of an equilateral triangle.*[2]

See Figure 4.2.

Paul Erdős (1913–1996) asserted that in some Platonic realm there is a book that contains the best proofs of mathematical theorems. In 1995 John H. Conway found a beautiful proof of Morley's theorem that may well be in Erdős' book. Conway's proof is elegant, memorable, and its diagram requires only six extra lines. Also, the proof proceeds by the neat method of starting with an equilateral triangle and working backwards to create a triangle similar to the given triangle, as in the proof of A Square inside Every Triangle.

Following Conway, let

$$\theta^+ = \theta + 60° \quad \text{and} \quad \theta^{++} = \theta + 120°,$$

where θ is an angle.

Suppose that our triangle has angles 3α, 3β, and 3γ, so that $\alpha + \beta + \gamma = 60°$. Start with an equilateral triangle of arbitrary size.

---

[2]Morley's theorem, a gem of geometry, was discovered in 1899 by Frank Morley (1860–1937). Morley worked in the areas of algebra and geometry. It isn't possible to construct the angle trisectors of an arbitrary angle using only straight-edge and compass. Is Morley's theorem a part of Euclidean geometry?

Next, form three triangles with angles shown in the figure below, and with the sides established by shaded segments congruent to the sides of the equilateral triangle.

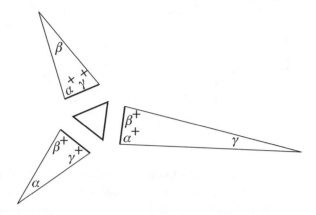

We can easily check that these angle measures really do give triangles. For example,

$$\alpha + \beta^+ + \gamma^+ = (\alpha + \beta + \gamma) + 60° + 60° = 60° + 60° + 60° = 180°.$$

The shaded edges determine the sizes of these triangles.

Finally, form three more triangles, with angles shown below.

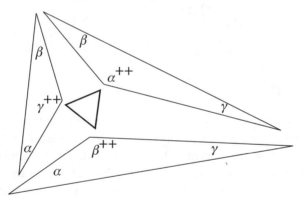

We need to say how large the triangles are, which we will do in a moment. For now, we check that the given angles make triangles. For example,

$$\alpha + \beta + \gamma^{++} = (\alpha + \beta + \gamma) + 120° = 180°.$$

To determine the sizes of the new triangles, drop pairs of equal line segments from a vertex of each triangle to the opposite side. Let's take the triangle with angles α, β, and γ^{++} as an example. Drop line segments from the vertex with angle γ^{++} to the opposite side so that the indicated angles are both γ^+. Let the length of the line segments be equal to the side of the equilateral triangle. This determines the size of this triangle. The sizes of the other two triangles are determined similarly. The appearance of the diagram would change slightly if one of the angles in the original triangle is a right angle or an obtuse angle. For example, if $3\gamma > 90°$, then $\gamma^+ > 90°$. How does this change the picture?

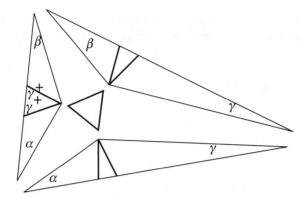

We claim that the seven triangles thus formed fit together to make a triangle similar to the given triangle. Since we have created a triangle similar to the given triangle for which the result is true, and the given triangle was arbitrary, we have proved the result for all triangles.

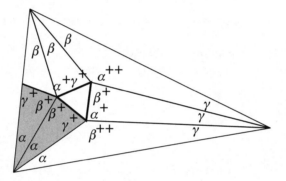

To prove that the triangles fit together, we must check that the four angles around each vertex of the equilateral triangle sum to 360°. I leave this as an exercise. We must also check that the sides fit together. The three triangles sharing a side with the equilateral triangle fit because the shared sides were formed to have length equal to the side of the equilateral triangle. To see that the other three triangles fit, notice that the two shaded triangles are congruent (because they have the same angles and a pair of congruent corresponding sides); hence the shaded triangles fit along a common side. Making this type of observation for five analogous cases completes the proof.

For other proofs of Morley's theorem, see [26]. There are many types of proofs, including one using complex numbers, but I believe that Conway's proof is the simplest.

4.3   The Euler Line

Every triangle has four well-known centers. The *orthocenter* H is the intersection of the three altitudes. The *centroid* G (the center of gravity of the triangle) is the intersection of the three medians. The *circumcenter* O (the center of the circumscribed circle) is the intersection of the perpendicular bisectors of the three sides. The *incenter* I (the center of the inscribed circle) is the intersection of the three angle bisectors.

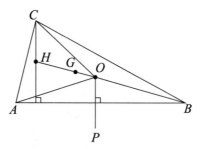

Figure 4.3. The Euler line (OGH).

Leonhard Euler (1707–1783) discovered that H, G, and O are collinear. The line that they lie on is called the *Euler line*. Furthermore, G lies one-third of the way along the Euler line from O to H. See Figure 4.3.

We will give a vector proof. In vector notation, we can describe Euler's discovery as follows.

Theorem. *If H, G, and O are the orthocenter, centroid, and circumcenter of a triangle, then*

$$3\overrightarrow{OG} = \overrightarrow{OH}.$$

The coordinates of the centroid are the averages of the coordinates of the three vertices of the triangle. This implies that

$$\overrightarrow{GA} + \overrightarrow{GB} + \overrightarrow{GC} = \overrightarrow{0}.$$

By the definition of vector addition,

$$\overrightarrow{OG} + \overrightarrow{GA} = \overrightarrow{OA}$$

$$\overrightarrow{OG} + \overrightarrow{GB} = \overrightarrow{OB}$$

$$\overrightarrow{OG} + \overrightarrow{GC} = \overrightarrow{OC}.$$

Adding, we get

$$3\overrightarrow{OG} = \overrightarrow{OA} + \overrightarrow{OB} + \overrightarrow{OC}.$$

We will show that the right side is equal to $\overrightarrow{OH}$.

The vector sum $\overrightarrow{OA} + \overrightarrow{OB}$ is a vector represented by the diagonal of the parallelogram spanned by $\overrightarrow{OA}$ and $\overrightarrow{OB}$. By definition of the circumcenter, these two vectors have the same length and the parallelogram they span is a rhombus. Since the diagonals of a rhombus are perpendicular, $\overrightarrow{OA} + \overrightarrow{OB}$ is a vector on the line OP, perpendicular to side AB. Hence, $\overrightarrow{OA} + \overrightarrow{OB}$ is parallel to the altitude from C. It follows by the definition of vector addition that $\overrightarrow{OA} + \overrightarrow{OB} + \overrightarrow{OC}$ is a vector from O to the altitude from C. A similar argument shows that $\overrightarrow{OA} + \overrightarrow{OB} + \overrightarrow{OC}$ is a vector from O to the altitudes from A and B. Since H is the intersection of the altitudes, $\overrightarrow{OA} + \overrightarrow{OB} + \overrightarrow{OC} = \overrightarrow{OH}$.

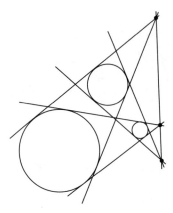

Figure 4.4. The circles, common external tangents, and collinear points of Monge's theorem.

4.4   Monge's Theorem

Theorem. *Given three circles in the plane, with different radii and none inside another, the three pairs of common external tangents of the circles intersect in three collinear points.*[3]

See Figure 4.4.

Suppose that the circles are C_1, C_2, C_3, with centers c_1, c_2, c_3, and radii r_1, r_2, r_3, respectively. Let p_{ij} be the intersection of the external tangents to C_i and C_j, where $1 \leq i < j \leq 3$. Observe (using similar triangles) that

$$p_{ij} = c_i + \frac{r_i}{r_i - r_j}(c_j - c_i) = \frac{r_i c_j - r_j c_i}{r_i - r_j}.$$

It follows that

$$r_1(r_2 - r_3)p_{23} + r_2(r_3 - r_1)p_{13} + r_3(r_1 - r_2)p_{12} = 0,$$

and therefore p_{12}, p_{23}, and p_{13} are collinear, since the scalar coefficients sum to 0.

4.5   Power Means

Power means are a grand generalization of the arithmetic mean and geometric mean. They have been studied for nearly 200 years but there are still interesting open questions concerning them, and some problems have been solved only recently.

A real-valued function f is *convex* on an interval I if

$$f((1 - \lambda)a + \lambda b) \leq (1 - \lambda)f(a) + \lambda f(b),$$

for all $a, b \in I$ and $0 \leq \lambda \leq 1$.

A straightforward application of calculus shows that if $f''(x) \geq 0$ for all $x \in I$, then f is convex on I. For example, the function $f(x) = -\ln x$ is convex on the interval $(0, \infty)$.

---

[3]This theorem is attributed to Gaspard Monge (1746–1818), the inventor of descriptive geometry.

If a function is convex, then it satisfies an inequality due to Johan Jensen (1859–1925):

Theorem (Jensen's Inequality). *Suppose that f is convex on I. If $a_1, \ldots, a_n \in I$ and $\lambda_1, \ldots, \lambda_n$ are nonnegative real numbers such that $\lambda_1 + \cdots + \lambda_n = 1$, then*

$$f\left(\sum_{i=1}^{n} \lambda_i a_i\right) \le \sum_{i=1}^{n} \lambda_i f(a_i).$$

The proof is by mathematical induction on n.

Suppose that $a_1, \ldots, a_n$ are positive numbers and $w_1, \ldots, w_n$ are positive numbers (weights) such that $\sum w_i = 1$. For $-\infty \le r \le \infty$, let

$$M_r = \begin{cases} \left(\sum_{i=1}^{n} w_i a_i^r\right)^{1/r} & -\infty < r < \infty, \ r \ne 0 \\[2mm] \prod_{i=1}^{n} a_i^{w_i} & r = 0 \\[2mm] \max\{a_i\} & r = \infty \\[2mm] \min\{a_i\} & r = -\infty. \end{cases}$$

We call M_r the r-th *power mean* of the numbers $a_1, \ldots, a_n$ with weights $w_1, \ldots, w_n$. We can make the values of the a_i explicit, if we wish, using the notation $M_r(a_1, a_2, \ldots, a_n)$.

The functions M_r with $r = -1, 0, 1$, and 2 are called, respectively, the *harmonic mean* (HM), *geometric mean* (GM), *arithmetic mean* (AM), and *quadratic mean* (QM):

$$M_{-1} = \frac{1}{\sum_{i=1}^{n} w_i/a_i},$$

$$M_0 = \prod_{i=1}^{n} a_i^{w_i},$$

$$M_1 = \sum_{i=1}^{n} w_i a_i,$$

$$M_2 = \left(\sum_{i=1}^{n} w_i a_i^2\right)^{1/2}.$$

Applying Jensen's inequality to the convex function $f(x) = -\ln x$ yields the arithmetic mean–geometric mean (AM–GM) inequality (with weights):

$$M_0 \le M_1.$$

The equal-weight case of the AM–GM inequality is often useful:

$$(a_1 \ldots a_n)^{1/n} \le \frac{a_1 + \cdots + a_n}{n},$$

where $a_1, \ldots, a_n$ are positive numbers. Equality holds if and only if all the a_i are equal.

Theorem (Power Means). *Let $a_1, \ldots, a_n, w_1, \ldots, w_n$ be fixed. Then M_r is a continuous and increasing function of r for $-\infty \leq r \leq \infty$. Moreover, M_r is a strictly increasing function of r unless all the a_i are equal.*

Here is a proof. For $0 < r < s < \infty$, let $t = s/r > 1$ and $f(x) = x^t$, for $x > 0$. We have $f''(x) = t(t-1)x^{t-2} \geq 0$, so f is a convex function. By Jensen's inequality,

$$f\left(\sum_{i=1}^{n} w_i a_i^r\right) \leq \sum_{i=1}^{n} w_i f(a_i^r).$$

Hence

$$\left(\sum_{i=1}^{n} w_i a_i^r\right)^{s/r} \leq \sum_{i=1}^{n} w_i a_i^s$$

and

$$\left(\sum_{i=1}^{n} w_i a_i^r\right)^{1/r} \leq \left(\sum_{i=1}^{n} w_i a_i^s\right)^{1/s}.$$

This shows that $M_r \leq M_s$.

For $0 < r < \infty$, we apply the AM–GM inequality to $a_1^r, \ldots, a_n^r$ and obtain

$$\left(\sum_{i=1}^{n} w_i a_i^r\right)^{1/r} \geq \prod_{i=1}^{n} a_i^{w_i}.$$

Hence $M_0 \leq M_r$.

The cases $-\infty < r < s \leq 0$ are covered by these results and the identity

$$M_{-r}(a_1, \ldots, a_n) = (M_r(1/a_1, \ldots, 1/a_n))^{-1}.$$

If $0 < r < \infty$, then

$$M_r \leq \left(\sum_{i=1}^{n} w_i \max\{a_i\}^r\right)^{1/r} = \max\{a_i\} = M_\infty.$$

Similarly, $M_{-\infty} \leq M_r$ if $-\infty < r < 0$. Therefore, M_r is an increasing function of r. It is easy to show that there is strict inequality unless $a_1 = a_2 = \cdots = a_n$.

Because M_r is a composition of continuous function, it is continuous on the intervals $(-\infty, 0)$ and $(0, \infty)$. To prove that M_r is continuous for all $r \in [-\infty, \infty]$, we must show that M_r is continuous at 0, ∞, and $-\infty$. To show that M_r is continuous at ∞, we apply the squeeze principle. We have already shown that $M_r \leq M_\infty$. Since

$$M_r \geq \left(w_i \cdot M_\infty^r\right)^{1/r} = (w_i)^{1/r} \cdot M_\infty$$

and $\lim_{r \to \infty} (w_i)^{1/r} = 1$, it follows that $\lim_{r \to \infty} M_r = M_\infty$. Hence M_r is continuous at ∞. A similar proof shows that M_r is continuous at $-\infty$. Finally, the continuity of M_r at $r = 0$ is proved by squeezing M_r between M_0 and a bound that tends to M_0 as $r \to 0$. We consider the case where r is positive (the case where r is negative follows from the identity above). We have already shown that $M_0 \le M_r$. From the AM–GM inequality,

$$\frac{1}{\sum w_i a_i^r} = \frac{\sum w_i a_i^r \left(\frac{1}{a_i^r}\right)}{\sum w_i a_i^r} \ge \prod_{i=1}^{n} \left(\frac{1}{a_i^r}\right)^{w_i a_i^r / \sum w_i a_i^r}.$$

Taking the reciprocal and the $1/r$-th power yields

$$M_r \le \left(\prod_{i=1}^{n} a_i^{w_i a_i^r}\right)^{1/\sum w_i a_i^r}.$$

Since the upper bound tends to M_0 as $r \to 0$, we conclude that M_r is continuous at 0.

We thus obtain the following chain of classical inequalities:

$$\min\{a_i\} \le \mathrm{HM} \le \mathrm{GM} \le \mathrm{AM} \le \mathrm{QM} \le \max\{a_i\}.$$

Graphing M_r for various values of the weights and variables may give the impression that the curve always has a single inflection point (where it changes from convex to concave). Often, the curve looks like the picture below.

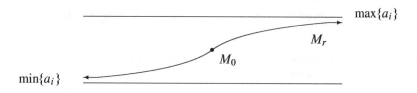

Since M_r has two horizontal asymptotes, the curve has at least one inflection point. But is there always exactly one?

Harold Shniad [47] found the counterexample

$$M_r = (0.1e^r + 0.8e^{2r} + 0.1e^{3r})^{1/r}.$$

(We have written the powers of the variables as equivalent exponential functions.) A computer algebra system shows that the second derivative of M_r changes sign three times, so there are three points of inflection: M_r'' is positive for $r = -2$, negative for $r = -1$, positive for $r = 0$, and negative for $r = 4$.

In 2008 Phan Thanh Nam and Mach Nguyet Minh [44] showed that for two variables, the function M_r is indeed convex-concave (having one inflection point). Their admirable proof deals with the complicated algebra involved in the second derivative. Is there is a simpler proof?

If there are more than two variables with equal weights, by setting some of the variables equal, Schniad's function can be written as a power mean with ten variables and equal weights. So, in this case, the curve has more than one inflection point.

The question of whether the power mean for more than two variables and equal weights is always convex–concave is, to the best of my knowledge, unsolved.

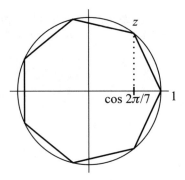

Figure 4.5. A regular heptagon.

4.6 Regular Heptagon

A regular polygon with n sides is constructible by straightedge and compass if and only if n is of the form

$$2^k p_1 \ldots p_m,$$

where $k, m \geq 0$ and the p_i are distinct primes of the form $2^{2^j} + 1$, with $j \geq 0$. Primes of this form are called *Fermat primes*, and the only ones known are 3, 5, 17, 257, and 65537, corresponding to $j = 0, 1, 2, 3$, and 4.

If we allow the use of an angle-trisecting device, then certain other regular polygons can be constructed.

Theorem. *A regular heptagon, which has seven sides, can be constructed using straightedge, compass, and an angle trisecting device.*

We will show that a regular (convex) heptagon can be constructed in the complex plane. The vertices are $1, z, z^2, z^3, z^4, z^5$, and z^6, where $z = e^{2\pi i/7}$. See Figure 4.5. The construction amounts to constructing a segment of length $z + z^6 = z + 1/z = 2\cos(2\pi/7)$, a real number, for this is twice the projection of z onto the real axis.

From the equation $z^7 = 1$ we have

$$z^6 + z^5 + z^4 + z^3 + z^2 + z + 1 = 0.$$

Letting $a = z + 1/z$ (a real number), we obtain

$$a^3 + a^2 - 2a - 1 = 0.$$

The irreducibility of this cubic polynomial is the reason why a regular heptagon is not constructible by straightedge and compass alone. A minimal polynomial of a constructible length must have a degree that is a power of 2.

To eliminate the coefficient of a^2, we make the substitution $a = (b - 1)/3$ and obtain

$$b^3 - 21b - 7 = 0.$$

We can solve this equation by using the formula for the cosine of a triple angle:

$$\cos 3\theta = 4\cos^3 \theta - 3\cos \theta.$$

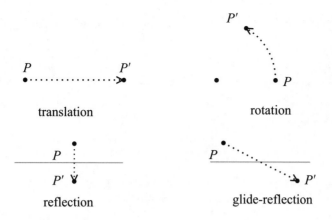

Figure 4.6. The four types of isometries of the Euclidean plane.

Our trisection procedure allows us to construct $\cos\theta$ when given $\cos 3\theta$. The change of variables $b = 2\sqrt{7}c$ puts our equation in the proper form.

Using straightedge and compass, we can construct any rational number and take square roots. Therefore, with the angle trisecting device we can construct a and thereby construct the regular heptagon.

See [20] for a discussion of the construction of the regular heptagon and the regular triskaidecagon, with thirteen sides, along with the following characterization of the regular polygons that can be constructed using straight-edge, compass, and angle trisecting device:

Theorem. *A regular n-gon can be constructed with straightedge, compass, and angle trisector if and only if n is of the form*

$$2^k 3^l p_1 \ldots p_m,$$

where $k, l, m \geq 0$, and the p_i are distinct primes greater than 3 of the form $2^h 3^j + 1$, with $h, j \geq 0$.

4.7 Isometries of the Plane

An *isometry* of the Euclidean plane $\mathbf{R}^2$ is a function $f : \mathbf{R}^2 \to \mathbf{R}^2$ that preserves distances:

$$|f(a) - f(b)| = |a - b|, \quad \text{for all } a, b \in \mathbf{R}^2.$$

The four types of isometries, as shown in Figure 4.6, are *translations*, *rotations*, *reflections*, and *glide-reflections*. A glide-reflection is a composition of a translation and a reflection in a line parallel to the direction of translation. Translations and glide-reflections have no fixed points (although a glide-reflection's reflecting line is fixed as a set of points), while rotations have one fixed point (the center) and reflections have a line of fixed points. Translations and rotations are orientation-preserving, while reflections and glide-reflections are orientation-reversing. Given that every isometry of the plane is one of these four types, we will prove that all isometries are given by two families of complex functions. For this purpose, we represent the Euclidean plane as $\mathbf{C}$.

Theorem. *Every isometry of the Euclidean plane is of the form*

$$f(z) = \alpha z + \beta \quad or \quad f(z) = \alpha \bar{z} + \beta, \quad where \; \alpha, \beta \in \mathbf{C}, \; |\alpha| = 1.$$

The first function is an orientation-preserving isometry; the second is an orientation-reversing one.

A translation is represented as

$$f(z) = z + \beta, \quad \beta \in \mathbf{C},$$

where β gives the direction and magnitude of the translation.

A rotation with center at the origin is represented as

$$f(z) = \alpha z, \quad \alpha \in \mathbf{C}, \; |\alpha| = 1.$$

The angle of the rotation is $\arg \alpha$.

A rotation with an arbitrary center is represented using conjugation (in the group theory sense: ghg^{-1}) of a rotation by a translation:

$$f(z) = \alpha(z - \gamma) + \gamma, \quad \alpha, \gamma \in \mathbf{C}, \; |\alpha| = 1.$$

The angle of rotation is $\arg \alpha$ and the center of rotation is γ.

If $f(z) = \alpha z + \beta$, with $\alpha, \beta \in \mathbf{C}$, $|\alpha| = 1$, then f is a translation or a rotation. Indeed, if $\alpha = 1$, then f is a translation, while if $\alpha \neq 1$, then f is the rotation given by

$$f(z) = \alpha \left(z - \frac{\beta}{1 - \alpha} \right) + \frac{\beta}{1 - \alpha}.$$

Reflection with respect to the x-axis is represented as

$$f(z) = \bar{z}.$$

Now let us consider a reflection with respect to a line through the origin. Suppose that the reflecting line is given by the complex number ω (as a vector), with $|\omega| = 1$. Then reflection with respect to this line is effected by conjugation:

$$f(z) = \omega \overline{(\omega^{-1} z)} = \omega^2 \bar{z}.$$

Hence $\alpha = \omega^2$.

Reflection with respect to a line parallel to ω is effected by conjugation by a translation $si\omega$, for some real s:

$$f(z) = \omega^2 \overline{(z - si\omega)} + si\omega = \omega^2 \bar{z} + 2si\omega, \quad \omega \in \mathbf{C}, \; |\omega| = 1, \; s \in \mathbf{R}.$$

Glide-reflection with respect to a line through the origin is represented as

$$f(z) = \omega^2 \overline{(z + t\omega)} = \omega^2 \bar{z} + t\omega, \quad \omega \in \mathbf{C}, \; |\omega| = 1, \; t \in \mathbf{R}.$$

Glide-reflection with respect to an arbitrary line is represented as

$$f(z) = \omega^2 \bar{z} + 2si\omega + t\omega, \quad \omega \in \mathbf{C}, \; |\omega| = 1, \; s, t \in \mathbf{R}.$$

The vector $2si\omega$ is perpendicular to ω and $t\omega$ is parallel to ω.

If $f(z) = \alpha\bar{z} + \beta$, with $\alpha, \beta \in \mathbf{C}$, $|\alpha| = 1$, then f is a reflection or a glide-reflection. For let $\alpha = \omega^2$, and $2si\omega$ and $t\omega$ be the perpendicular and parallel components of β with respect to ω, respectively. Then we may write

$$f(z) = \omega^2\bar{z} + 2si\omega + t\omega,$$

and we see that f is a glide-reflection (or reflection, if $t = 0$).

It is easy to show that the set of isometries comprise a group under composition. They form a closed set with an identity and inverses, and composition of functions is associative. The group of isometries of the Euclidean plane is generated by one type of isometry: reflections.

- A translation is the composition of two reflections.

- A rotation is the composition of two reflections.

- A reflection is one reflection.

- A glide-reflection is the composition of a reflection and a translation, so it is equivalent to three reflections.

Thus, every isometry in the Euclidean plane is a composition of one, two, or three reflections.

4.8   Symmetries of Regular Convex Polyhedra

A *regular polyhedron* has the property that under symmetry all its vertices are equivalent, all its edges are equivalent, and all its faces are equivalent. There are five regular convex polyhedra: tetrahedron, cube, octahedron, dodecahedron, and icosahedron. See Figure 4.7. A simple proof uses Euler's formula, $V - E + F = 2$, where E is the number of edges, V the number of vertices, and F the number of faces of the polyhedron. See, e.g., [26].

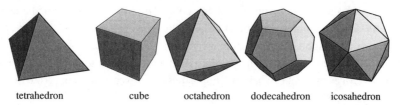

tetrahedron   cube   octahedron   dodecahedron   icosahedron

Figure 4.7. The five regular convex polyhedra.

	vertices	edges	faces	edges per vertex	edges per face
tetrahedron	4	6	4	3	3
cube	8	12	6	3	4
octahedron	6	12	8	4	3
dodecahedron	20	30	12	3	5
icosahedron	12	30	20	5	3

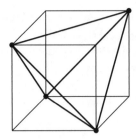

Figure 4.8. A regular tetrahedron inscribed in a cube.

A *symmetry* of a polyhedron is a motion that moves it so that it occupies its original space. The set of all symmetries forms a group under composition of motions. We are considering only proper symmetries of the polyhedron, those that preserve orientation. Reflections are excluded. What are the symmetry groups of the regular convex polyhedra?

The *dual polyhedron* of a polyhedron is the polyhedron obtained by putting a vertex at the center of each face of the given polyhedron and joining two new vertices if the faces of the given polyhedron share an edge. A polyhedron and its dual have the same symmetry group. The cube and the octahedron are duals, so they have the same symmetry group. The icosahedron and dodecahedron are duals, so they have the same symmetry group. The tetrahedron is self-dual.

If we pick up a cube and set it down again so that it occupies its original space, then its vertices, edges, and faces of may have changed position. The symmetry group of the cube is the group of all such ways to reposition the cube. It's easy to find the order (the number of elements) of the symmetry group of the cube. We can set the cube down on any of its six faces, and then rotate it in any of four ways. Hence there are $6 \cdot 4 = 24$ symmetries. However, we still need to decide which 24-element group this is.

We know that the symmetric group S_4, the group of permutations on 4 objects, has $4! = 24$ elements. In fact, the symmetry group of the cube, and of the regular octahedron, is isomorphic to S_4. To see this, we need only show that the cube contains some four elements that are permuted in all ways by the symmetries of the cube, and that each permutation of them gives rise to a unique symmetry of the cube. The four diagonals of the cube have this property. Every symmetry of the cube permutes the diagonals, and conversely every permutation of the diagonals comes from a symmetry of the cube.

To find the symmetry group of the regular tetrahedron, place the vertices of the tetrahedron at four vertices of a cube. These are opposite pairs of vertices on opposite faces of the cube. See Figure 4.8.

Now that we know the symmetry group of the cube (S_4), we find that the symmetry group of the regular tetrahedron goes along for the ride. Every symmetry of the cube automatically gives a symmetry of the regular tetrahedron (with the tetrahedron inscribed in the cube), or it moves the vertices of the tetrahedron to the other four vertices of the cube. How many symmetries of the regular tetrahedron are there? Since we can put the tetrahedron down on any of its four faces and then rotate it in any of three ways, the regular tetrahedron has $4 \cdot 3 = 12$ symmetries. The symmetries of the tetrahedron comprise a subgroup of S_4 of order 12. It can be shown that this subgroup is the alternating group A_4, consisting of even permutations in S_4.

Similar arguments show that the symmetry group of both the regular icosahedron and the regular dodecahedron is the alternating group A_5. A cube can be inscribed in a regular dodecahedron so that each of the twelve edges of the cube is a diagonal of a face of the dodecahedron. See the diagram below. Each face of the dodecahedron has five diagonals (the edges of a pentagram). Hence, five such cubes can be inscribed. Symmetries of the dodecahedron permute the five cubes. It follows that the symmetry group of the dodecahedron is a subgroup of the symmetric group S_5. Since the symmetry group has order 60 (why?), it is A_5.

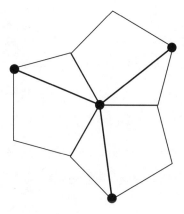

Theorem. *The regular tetrahedron has symmetry group A_4. The cube and regular octahedron have symmetry group S_4. The regular dodecahedron and regular icosahedron have symmetry group A_5.*

See [13] for a panoramic and detailed survey of symmetries of geometric figures.

4.9 Polynomial Symmetries

A *symmetry* of a polynomial (in several variables) is a permutation of the polynomial's variables that leaves the polynomial unchanged. For example, the polynomial $xy^2 + yz^2 + zx^2$ has symmetry group $\mathbf{Z}_3$, since the permutations of the variables that preserve the polynomial are $(x)(y)(z)$, (x, y, z), and (z, y, x).

Theorem. *Given a finite group G of order n, there exists a polynomial in n variables, all of whose coefficients are 1, with symmetry group G.*

Given any n-element group G a group of permutations of the set $\{1, \ldots, n\}$, we see by inspection that the polynomial

$$f(x_1, x_2, \ldots, x_n) = \sum_{\sigma \in G} x_{\sigma(1)} x_{\sigma(2)}^2 \ldots x_{\sigma(n)}^n$$

has symmetry group G. The monomial $x_1 x_2^2 \ldots x_n^n$ is mapped by σ to another monomial $x_{\sigma(1)} x_{\sigma(2)}^2 \ldots x_{\sigma(n)}^n$ in the polynomial if and only if $\sigma \in G$.

If we do this construction for $\mathbf{Z}_3$, we obtain

$$f(x_1, x_2, x_3) = x_1 x_2^2 x_3^3 + x_2 x_3^2 x_1^3 + x_3 x_1^2 x_2^3.$$

Factoring out $x_1 x_2 x_3$, and making the change of variables $x_1 \leftarrow x$, $x_2 \leftarrow y$, and $x_3 \leftarrow z$, we obtain our polynomial $xy^2 + yz^2 + zx^2$.

The *quaternion group* Q consists of eight elements, ± 1, $\pm i$, $\pm j$, and $\pm k$, that satisfy the rules $i^2 = j^2 = k^2 = -1$ and $ij = k$, $jk = i$, $ki = j$.

Let's find a polynomial whose symmetry group is the quaternion group. We use the labeling

$$\begin{array}{cccccccc} 1 & 2 & 3 & 4 & 5 & 6 & 7 & 8 \\ 1 & -1 & i & -i & j & -j & k & -k. \end{array}$$

Using the multiplication rules, we find that the polynomial is

$$x_1 x_2^2 x_3^3 x_4^4 x_5^5 x_6^6 x_7^7 x_8^8 + x_2 x_1^2 x_4^3 x_3^4 x_6^5 x_5^6 x_8^7 x_7^8$$

$$+ x_3 x_4^2 x_2^3 x_1^4 x_7^5 x_8^6 x_6^7 x_5^8 + x_4 x_3^2 x_1^3 x_2^4 x_8^5 x_7^6 x_5^7 x_6^8$$

$$+ x_5 x_6^2 x_8^3 x_7^4 x_2^5 x_1^6 x_3^7 x_4^8 + x_6 x_5^2 x_7^3 x_8^4 x_1^5 x_2^6 x_4^7 x_3^8$$

$$+ x_7 x_8^2 x_5^3 x_6^4 x_4^5 x_3^6 x_2^7 x_1^8 + x_8 x_7^2 x_6^3 x_5^4 x_3^5 x_4^6 x_1^7 x_2^8.$$

It has symmetry group Q.

A finite group G has many realizations. We have seen that G is the symmetry group of a polynomial. Some other realizations are:

- G is given by its multiplication table.

- G is isomorphic to a set of permutations, under composition. (Arthur Cayley[4])

- G is isomorphic to a matrix group. This is called a group *representation*.

- G is the automorphism group of a finite graph. (Roberto Frucht[5])

- G is the automorphism group of a compact Riemann surface. (Adolf Hurwitz[6])

- G is the automorphism group of a perfect binary code. (Kevin Phelps[7])

- G is the automorphism group of a distributive lattice. (Garrett Birkhoff[8])

- G is given by a presentation. An example of a group presentation is given in Appendix A.

It is not known whether the following is true:

- G is the automorphism group of an algebraic extension of the rational numbers.

---

[4] Arthur Cayley (1821–1895) was a pioneer in the areas of algebra, non-Euclidean geometry, and combinatorics.

[5] Roberto Frucht (1906–1997) was a graph theorist.

[6] Adolf Hurwitz (1859–1919) was an algebraist, geometer, and number theorist.

[7] Kevin Phelps is a researcher in the areas of coding theory, combinatorics, and graph theory.

[8] Garrett Birkhoff (1911–1996) was an algebraist, working specifically in the areas of lattice theory and universal algebra.

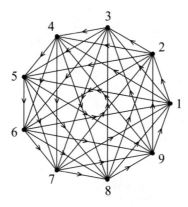

Figure 4.9. A tournament on nine vertices.

4.10   Kings and Serfs

A *tournament* is a complete finite graph in which each edge has been replaced by a directed arrow. Figure 4.9 shows a tournament on nine vertices.

In a tournament, a *King* is a vertex from which every other vertex can be reached in one or two steps. A *Serf* is a vertex that can be reached from every other vertex in one or two steps. Every vertex in the tournament of Figure 4.9 is both a King and a Serf. We will show that this is typical.

The *outdegree* of a vertex v is the number of directed edges that emanate from v. The *indegree* of v is the number of edges directed to v. Every vertex in the tournament of Figure 4.9 has outdegree 4 and indegree 4.

Theorem. *(H. G. Landau). Every tournament has a King.*

Consider a vertex v of maximum outdegree. We will prove that v is a King of the tournament. Suppose that there are edges directed away from v to r vertices, $u_1, \ldots, u_r$. Suppose also that there is a vertex w that cannot be reached in one or two steps from v. Then w is not among the u_i and there are edges directed from w to all the u_i and to v. But this means that the outdegree of w is at least $r + 1$, contradicting the choice of v.

The assertion that every tournament has a Serf is the dual statement of Landau's theorem.

In a random tournament, the direction of each edge is chosen randomly with equal probability of going in either direction. In a large random tournament, almost assuredly every vertex is both a King and a Serf.

Theorem. *(Stephen B. Maurer). In a random tournament on n vertices, the probability that every vertex is a King and a Serf tends to 1 as n tends to infinity.*

A tournament lacks the desired property if and only if there exists a pair of vertices v_1 and v_2 with $v_1 \leftarrow v_2$ such that there is no path of length 2 from v_1 to v_2. In a random tournament, this happens with probability at most

$$\binom{n}{2} \left(\frac{3}{4} \right)^{n-2}.$$

The reason is that there are $\binom{n}{2}$ choices for the "bad" vertices v_1 and v_2, and $n - 2$ choices for a third vertex w; the probability that there is a path from v_1 to w to v_2 is $(1/2)^2 = 1/4$.

Hence, the probability that there is no path of length 2 from v_1 to v_2 is $(3/4)^{n-2}$. Therefore, since probabilities are subadditive, the probability that there exist such v_1 and v_2, with no path of length 2 from v_1 to v_2, is bounded by the number of choices of v_1 and v_2 times $(3/4)^{n-2}$.

Our upper bound is the product of a polynomial and an exponential function with base less than 1. As $n \to \infty$, the exponential function dominates and the product tends to 0. Hence, the probability of the complementary event—the event that every vertex is a King and a Serf—tends to 1 as n tends to ∞.

4.11   The Erdős--Szekeres Theorem

In any sequence of ten distinct real numbers, there exists an increasing subsequence of four terms or a decreasing subsequence of four terms. The terms in the subsequence need not be consecutive.

For example, the ten-term sequence

$$7, \ 8, \ 4, \ 9, \ 5, \ 1, \ 6. \ 2, \ 3, \ 10$$

contains, among others, the four-term increasing subsequence 4, 5, 6, 10.

This assertion is an instance of the Erdős–Szekeres theorem, due to Paul Erdős (1913–1996) and George Szekeres (1911–2005), which we state below.

Let the terms of the sequence be $x_1, x_2, \ldots, x_{10}$. To each x_i, we associate an ordered pair (u_i, d_i), where u_i is the length of a longest increasing subsequence that begins with x_i (the u stands for "up"), and d_i is the length of a longest decreasing subsequence that begins with x_i (the d stands for "down"). Assume that the sequence doesn't contain an increasing or decreasing subsequence of length four. Then, for each i, we have $1 \leq u_i, d_i \leq 3$. Now we can invoke the famous *pigeonhole principle*, a staple of combinatorial mathematics.

Pigeonhole Principle.   If $N + 1$ objects are placed in N pigeonholes, then one of the pigeonholes contains at least two objects.

The proof is by contradiction.

We apply the pigeonhole principle with $N = 9$, where the objects are the ten numbers $1, 2, \ldots, 10$, and the pigeonholes are the nine ordered pairs $(0, 0), (0, 1), \ldots, (3, 3)$. For $1 \leq i \leq 10$, we place i in the pigeonhole corresponding to the ordered pair (u_i, d_i). The pigeonhole principle guarantees that some two numbers i and j, with $i < j$, are in the same pigeonhole; that is, some (u_i, d_i) and (u_j, d_j) are equal. But this is impossible, for if $x_i < x_j$ then $u_i > u_j$, while if $x_i > x_j$ then $d_i > d_j$. The contradiction means that there exists an increasing subsequence or a decreasing subsequence of length four.

Here is the general statement of the theorem.

Erdős–Szekeres Theorem.   *In any sequence of $mn + 1$ distinct real numbers, where m and n are positive integers, there exists an increasing subsequence of length $m + 1$ or a decreasing subsequence of length $n + 1$.*

The proof uses the pigeonhole principle as in the special case $m = n = 3$. The result of the theorem doesn't hold if we replace $mn + 1$ by mn. We can see this in our special case.

Remove the 10 at the end of the example sequence above, and there is no increasing or decreasing subsequence of length four.

How many sequences consisting of the numbers 1, 2, ..., 9, in some order, contain no monotonic (increasing or decreasing) subsequence of length 4? A computer search shows that there are 1764. This is interesting, because 1764 is a perfect square: $1764 = 42^2$. There are 42 fillings of a 3×3 array with the numbers 1 through 9, so that the numbers increase in each row and column. These are called *standard fillings*. Here is an example of a standard filling.

1	2	4
3	5	8
6	7	9

The number of standard fillings is given by the *hook length formula*. The *hook length* of a cell in a grid is the number of squares to the right and below that cell, plus one to count the cell itself. The hook lengths for the cells of a 3×3 grid are shown below.

5	4	3
4	3	2
3	2	1

The number of standard fillings of the 3×3 grid is the number of permutations of nine elements divided by the product of the hook lengths:

$$\frac{9!}{5 \cdot 4 \cdot 4 \cdot 3 \cdot 3 \cdot 3 \cdot 2 \cdot 2 \cdot 1} = 42.$$

Thus, the number of permutations of nine elements that contain no monotonic subsequence of length four is the square of the number of standard fillings of a 3×3 grid. We have touched on the rich theory of *Young tableaux*. See, e.g., [32].

4.12 Minkowski's Theorem

A region in the plane is *convex* if it contains the line segment joining any two of its points. The region is *centrally symmetric* if it contains the point $(-x, -y)$ whenever it contains the point (x, y). A *lattice point* in the plane is a point with integer coordinates. A convex centrally symmetric planar region certainly contains at least one lattice point: the origin. A famous theorem of Hermann Minkowski[9] asserts that if such a region has area greater than 4, then it must contain another lattice point.

---

[9] Hermann Minkowski (1864–1909) made fundamental contributions in number theory and relativity theory.

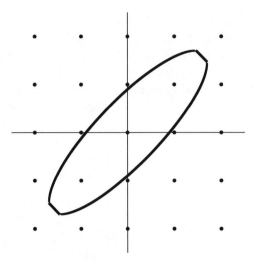

Figure 4.10. A planar region satisfying the conditions of Minkowski's theorem.

Minkowski's Theorem. *A convex centrally symmetric planar region of area greater than* 4 *contains a lattice point other than the origin.*

Figure 4.10 shows an example of a convex centrally symmetric region of area greater than 4. We see that it contains the lattice points $(1, 1)$ and $(-1, -1)$, in addition to $(0, 0)$.

It's a good exercise to show why the conclusion of Minkowski's theorem fails if the region isn't convex or isn't centrally symmetric.

We sketch a proof of Minkowski's theorem. Suppose that K is a convex centrally symmetric planar region of area greater than 4. Then we claim that K contains distinct points $v = (a, b)$ and $w = (c, d)$ such that $v - w$ is an ordered pair of even integers. The proof of this claim uses the pigeonhole principle, but in a different version from the one in the previous section. For each element of K, add or subtract an ordered pair of even integers so that the result is a pair of numbers both between -1 and 1. Since the area of K is greater than 4, some distinct points v and w are mapped to the same point, and hence have the desired property.

Since K is centrally symmetric, $-w$ is an element of K. Because K is convex, the midpoint of v and $-w$, that is,

$$\frac{v + (-w)}{2},$$

is in K. Since each coordinate of the numerator is an even integer, the midpoint is a lattice point, and it is not $(0, 0)$ since $v \neq w$. So, we have proved the existence of a lattice point in K other than the origin. Because of central symmetry, K contains two such lattice points.

We can deduce a consequence of Minkowski's theorem in number theory.

Theorem. *If p is a prime number of the form $4n + 1$, then p is the sum of two squares of integers: $p = x^2 + y^2$.*

For instance, $29 = 5^2 + 2^2$. The representation of p as a sum of two squares is unique up to the order of the squares, but we won't prove this. There is no representation of a prime p as a sum of two squares if $p \equiv 3 \pmod 4$. Squares modulo 4 are either 0 or 1, so the sum of two squares must be 0, 1, or 2.

Our proof requires a generalization of Minkowski's theorem to arbitrary planar lattices. Let v_1 and v_2 be linearly independent vectors in the plane. Then v_1 and v_2 generate a lattice Λ consisting of all sums of integral multiples of the two vectors:

$$\Lambda = \{m_1 v_1 + m_2 v_2 : m_1, m_2 \text{ integers}\}.$$

Let Δ be the area of the parallelogram spanned by v_1 and v_2.

Minkowski's Theorem (for an Arbitrary Lattice). *A convex centrally symmetric planar region of area greater than* 4Δ *contains a lattice point other than the origin.*

Let p be a prime of the form $4n + 1$. An important fact of number theory is that -1 is a square modulo p. A good way to think about this is in terms of the group of units (the nonzero elements) modulo p. For any prime, the units form a cyclic group; that is, the group consists of powers of an element called a *generator*. For instance, the group of units modulo 7 is generated by 3, because modulo 7 we have $3, 3^2 \equiv 2, 3^3 \equiv 6, 3^4 \equiv 4, 3^5 \equiv 5$, and $3^6 \equiv 1$. For our prime p, let g be a generator. Then $(g^n)^2 = g^{2n} \equiv -1 \pmod{p}$.

Let $h = g^n$ (whose square is -1), and define Λ by the vectors $v_1 = (h, 1)$ and $v_2 = (p, 0)$. The parallelogram spanned by v_1 and v_2 has area $\Delta = 1 \cdot p = p$.

If $(x, y) \in \Lambda$, then $x^2 + y^2 \equiv 0 \pmod{p}$. The reason is that $x = m_1 h + m_2 p$ and $y = m_1$, for some integers m_1 and m_2, and hence

$$x^2 + y^2 = (m_1 h + m_2 p)^2 + m_1^2 \equiv m_1^2(h^2 + 1) \equiv 0 \pmod{p}.$$

Let K be the open disk of radius $\sqrt{2p}$ centered at the origin. The area of K is $2\pi p > 4\Delta$, and of course a disk is convex and centrally symmetric. Minkowski's theorem guarantees the existence of a lattice point (x, y) in K other than the origin. For this lattice point,

$$0 < x^2 + y^2 < 2p.$$

Since $x^2 + y^2 \equiv 0 \pmod{p}$, we have $x^2 + y^2 = p$, a representation of p as a sum of two squares.

See [39] for an excellent introduction to Minkowski's theorem and its consequences.

4.13 Lagrange's Theorem

Lagrange's Theorem. *Every positive integer is a sum of four squares.*

For instance,

$$15 = 3^2 + 2^2 + 1^2 + 1^2.$$

Three squares do not always suffice; e.g., 7 isn't a sum of three squares.

Our proof of this number theory gem uses Minkowski's theorem. Let's start with the easiest case:

$$1 = 1^2 + 0^2 + 0^2 + 0^2.$$

We must show that every integer greater than 1 is a sum of four squares.

If m is a sum of four squares and n is a sum of four squares, then the product mn is a sum of four squares. To understand this, we turn—surprisingly—to alternative number systems.

The length of the complex number $z = a + bi$ is $|z| = \sqrt{a^2 + b^2}$. Let $z_1 = a + bi$ and $z_2 = c + di$. Then from the identity

$$|z_1||z_2| = |z_1 z_2|,$$

we obtain, upon squaring,

$$(a^2 + b^2)(c^2 + d^2) = (ac - bd)^2 + (ad + bc)^2.$$

Thus, the product of two sums of squares is itself a sum of squares. We can get the same result for sums of four squares by using quaternions (see Polynomial Symmetries). The length of a quaternion $q = a + bi + cj + dk$ is

$$|q| = \sqrt{a^2 + b^2 + c^2 + d^2}.$$

Let $q_1 = a + bi + cj + dk$ and $q_2 = w + xi + yj + zk$. Then the identity

$$|q_1||q_2| = |q_1 q_2|$$

yields

$$(a^2 + b^2 + c^2 + d^2)(w^2 + x^2 + y^2 + z^2)$$
$$= (aw - bx - cy - dz)^2 + (ax + bw + cz - dy)^2$$
$$+ (ay - bz + cw + dx)^2 + (az + by - cx + dw)^2.$$

Hence, a product of two sums of four squares is a sum of four squares. Since every integer greater than 1 factors into prime numbers, we need only prove that every prime is a sum of four squares.

We will use a generalization of Minkowski's theorem to arbitrary lattices in d-dimensional space. A *lattice* Λ in d-dimensional space is the collection of integer linear combinations of d independent vectors, $v_1, \ldots, v_d$:

$$\Lambda = \{m_1 v_1 + \cdots + m_d v_d : m_1, \ldots, m_d \text{ integers}\}.$$

Let Δ be the volume of the parallelepiped spanned by $v_1, \ldots, v_d$.

Minkowski's Theorem (for an Arbitrary Lattice in d-Dimensional Space). *A convex centrally symmetric region in d-dimensional space of volume greater than $2^d \Delta$ contains a lattice point other than the origin.*

Let p be a prime. For $p = 2$, we have $2 = 1^2 + 1^2 + 0^2 + 0^2$. So, let's assume that p is an odd prime. In order to carry out our proof that p is a sum of four squares, we will describe a lattice in 4-dimensional space. Its definition depends on the solution to a congruence modulo p.

We claim that there exist integers a and b such that

$$a^2 + b^2 \equiv -1 \pmod{p}.$$

There are exactly $(p + 1)/2$ distinct squares modulo p, namely, the squares modulo p of the first $(p + 1)/2$ integers. Thus, the quantities a^2 and $-(b^2 + 1)$ both take exactly

$(p + 1)/2$ values modulo p. By the pigeonhole principle, there exist a and b for which they are equal, so they satisfy the congruence.

We define the lattice Λ in 4-dimensional space to be the set of integer linear combinations of the vectors

$$v_1 = (p, 0, 0, 0)$$

$$v_2 = (0, p, 0, 0)$$

$$v_3 = (a, b, 1, 0)$$

$$v_4 = (b, -a, 0, 1).$$

Recall from Chapter 1 that a $d \times d$ determinant is equal to the (signed) volume of the parallelepiped in d-dimensional space spanned by its rows. Hence

$$\Delta = \begin{vmatrix} p & 0 & 0 & 0 \\ 0 & p & 0 & 0 \\ a & b & 1 & 0 \\ b & -a & 0 & 1 \end{vmatrix}.$$

This lower-triangular determinant is equal to the product of its diagonal entries: $\Delta = p^2$.

If $(x_1, x_2, x_3, x_4) \in \Lambda$, then there exist integers m_1, m_2, m_3, and m_4 such that

$$(x_1, x_2, x_3, x_4) = m_1(p, 0, 0, 0) + m_2(0, p, 0, 0) + m_3(a, b, 1, 0) + m_4(b, -a, 0, 1)$$

$$= (m_1 p + m_3 a + m_4 b, m_2 p + m_3 b - m_4 a, m_3, m_4).$$

Modulo p we have

$$x_1^2 + x_2^2 + x_3^2 + x_4^2 \equiv (m_3 a + m_4 b)^2 + (m_3 b - m_4 a)^2 + m_3^2 + m_4^2$$

$$\equiv m_3 a^2 + m_4 b^2 + m_3 b^2 + m_4 a^2 + m_3^2 + m_4^2$$

$$\equiv (a^2 + b^2 + 1)(m_3^2 + m_4^2)$$

$$\equiv 0.$$

Let K be the open four-dimensional hypersphere of radius $\sqrt{2p}$ centered at the origin. The formula for the volume of a d-dimensional hypersphere (ball) of radius r was given in Chapter 3:

$$\frac{\pi^{d/2} r^d}{(d/2)!}.$$

For K, we have $d = 4$ and $r = \sqrt{2p}$, so its volume is

$$2p^2 \pi^2.$$

Since $\pi^2 > 8$, this expression is greater than $16p^2$, and Minkowski's theorem implies the existence of a lattice point in K other than the origin. If (x_1, x_2, x_3, x_4) is such a point, then

$$0 < x_1^2 + x_2^2 + x_3^2 + x_4^2 < 2p,$$

and therefore $x_1^2 + x_2^2 + x_3^2 + x_4^2 = p$, a representation of p as a sum of four squares. This concludes the proof of Lagrange's theorem.

4.14   Van der Waerden's Theorem

An l-term *arithmetic progression* (l-AP) is a sequence

$$a, \; a + d, \; a + 2d, \; \ldots, \; a + (l - 1)d,$$

where a is the *initial term* of the sequence and d is the *common difference* between two consecutive terms of the sequence. For instance,

$$10, \; 15, \; 20, \; 25, \; 30, \; 35, \; 40$$

is a 7-AP with initial term 10 and common difference 5.

In 1927 B. L. van der Waerden[10] showed that if the set of positive integers **N** is partitioned into two classes, then at least one of the classes contains arbitrarily long arithmetic progressions.

Van der Waerden's Theorem. *If the set of positive integers (**N**) is partitioned into two classes, then at least one contains an l-AP for every $l \geq 1$.*

The theorem doesn't say that some class contains an infinite arithmetic progression. This isn't guaranteed, as you can see by defining the classes so that 1 is in the first class, 2 and 3 are in the second class, 4, 5, and 6 are in the first class, 7, 8, 9, and 10 are in the second class, and so on. Neither class contains an infinite arithmetic progression.

To prove van der Waerden's theorem, we will generalize it to allow for a partitioning of **N** into any finite number of classes.

Van der Waerden's Theorem (Infinite Version). *Let c be an integer greater than* 1. *If the set of positive integers (**N**) is partitioned into c classes, then at least one of the classes contains an l-AP for every $l \geq 1$.*

We call this theorem the "infinite version" because another version of van der Waerden's theorem concerns only finite sections of the positive integers. This statement is referred to as the "finite version" of the theorem.

We set $\mathbf{N}(n) = \{1, 2, 3, \ldots, n\}$.

Van der Waerden's Theorem (Finite Version). *Given integers $c \geq 1$ and $l \geq 1$, there exists a least integer $W(c, l)$ with the property that if $\mathbf{N}(W(c, l))$ is partitioned into c classes, then one of the classes contains an l-AP.*

We sometimes refer to the classes as *colors* and to the partition of $\mathbf{N}(W(c, l))$ as a c-*coloring*. An l-AP contained in a single class is called a *monochromatic l-AP*.

The values of $W(c, l)$ are called *van der Waerden numbers*.

The generalization to c colors is equivalent to the restriction to two colors. For in a coloring with c colors, all the colors but one could be combined to make one color, and hence produce a coloring using only two colors. If the monochromatic l-AP is found in the color that isn't a combined color, then we are done; otherwise, we can repeat the argument until we have a monochromatic l-AP in one of the c colors.[11]

---

[10]Bartel Leendert van der Waerden (1903–1996) made contributions in algebra and wrote a famous textbook on the subject called *Modern Algebra*.

[11]In proving the theorem that bears his name, van der Waerden worked with Emil Artin (1898–1962) and Otto Schreier (1901–1929). Schreier suggested the finite version of the theorem, while Artin suggested the generalization to arbitrarily many colors.

The finite version of van der Waerden's theorem is equivalent to the infinite version. To see that the finite version implies the infinite version, suppose that we have a partition of $\mathbf{N}$ into c classes. We want to show that one of the classes contains arbitrarily long finite arithmetic progressions. For any l, there exists an integer $W(c, l)$ such that no matter how $\mathbf{N}(W(c, l))$ is partitioned into c classes, one class contains a monochromatic l-AP. Thus, one class in the partition of $\mathbf{N}$ contains a monochromatic l-AP for each $l \geq 1$. Since there are only finitely many classes, one of the classes must contain monochromatic l-APs for infinitely many values of l. This class contains arbitrarily long monochromatic arithmetic progressions.

We will show that the infinite version implies the finite version for $c = 2$ (the general case is handled in the same way). Actually, we will prove the contrapositive; that is, we will assume that the finite version is false and show that the infinite version is false. That the finite version is false for $c = 2$ means that there is a positive integer l such that for every positive integer n there is a partition of $\mathbf{N}(n)$ into two classes neither of which contains an l-AP. For $n = 1, 2, 3, \ldots$, let $D(n)$ be a partition of $\mathbf{N}(n)$ into two classes, A and B, such that neither A nor B contains an l-AP. For instance, if $l = 3$ and $n = 6$, we can take $D(6)$ to be $\mathbf{N}(6) = \{1, 3, 4, 6\} \cup \{2, 5\}$, as neither subset contains a 3-AP. Consider the sequence S of such partitions:

$$S = \{D(1), \ D(2), \ D(3), \ \ldots\}.$$

The integer 1 occurs in one of the classes in each partition, so it must occur in the same class, either A or B, in infinitely many of them. Let

$$S(1) = \{D(1, 1), \ D(1, 2), \ D(1, 3), \ \ldots\}$$

be a subsequence of S in which 1 occurs in the same class in each partition. In the same way, we may form a subsequence $S(2)$ of $S(1)$ in which the integer 1 occurs in the same class and the integer 2 occurs in the same class (although 1 and 2 need not occur in the same class):

$$S(2) = \{D(2, 2), \ D(2, 2), \ D(2, 3), \ \ldots\}.$$

Continuing, we can get a sequence

$$S(n) = \{D(n, n), \ D(n, n + 1), \ D(n, n + 2), \ \ldots\},$$

in which the integer i occurs in the same class in each partition, for $i = 1, \ldots, n$.

Now we define a partition D of $\mathbf{N} = \{1, 2, 3, \ldots\}$ by putting m into the class in which it appears in $D(m, m)$. If the infinite version were true, then the partition D would have a class that contained an l-AP in the first k integers, for some k. But then $D(k, k)$ would contain an l-AP, which contradicts the fact that $D(k, k)$ was constructed so that it doesn't contain an l-AP. Hence the infinite version must be false. The contrapositive of this implication is that the infinite version implies the finite version. This is what we wanted to show.

We will now prove the finite version of van der Waerden's theorem. The proof is by mathematical induction on l, the number of terms in the monochromatic arithmetic progression. The result is trivially true for $l = 1$, since $W(c, 1) = 1$ for all c. It is also trivially true for $l = 2$, since $W(c, 2) = c + 1$ for all c. This statement is the basis of the induction.

Given $l \geq 2$, we assume that $W(c, l)$ exists for all $c \geq 2$, and we will prove the existence of $W(c, l + 1)$ for all $c \geq 2$. This will complete the induction.

We claim that $W(c, l + 1)$ exists and satisfies $W(c, l + 1) \leq f(c)$, where f is defined recursively:

$$f(1) = 2W(c, l)$$
$$f(n) = 2W(c^{f(n-1)}, l) f(n - 1), \quad n \geq 2.$$

Suppose that $\mathbf{N}(f(c))$, which we call a *c-block*, is c-colored without a monochromatic $(l + 1)$-AP, and $\mathbf{N}(f(c))$ is partitioned into $f(c)/f(c - 1)$ blocks of $f(c - 1)$ consecutive integers. Let's call these blocks of integers $(c - 1)$-*blocks*. Likewise, each $(c - 1)$-block is partitioned into $f(c - 1)/f(c - 2)$ blocks of $f(c - 2)$ consecutive integers, which we call $(c - 2)$-*blocks*. This partitioning takes place at each of the c levels, until each 1-block is partitioned into $2W(c, l)$ 0-*blocks* (which are integers).

By the definition of $W(c, l)$, the first half of each 1-block contains a monochromatic l-AP. The coloring of the elements of a 1-block induces a coloring of the 1-block itself: we assign one of $c^{f(1)}$ colors to the 1-block according to the way its elements are c-colored. Since $f(2) = 2W(c^{f(1)}, l) f(1)$, each 2-block contains $2W(c^{f(1)}, l)$ 1-blocks, so that by definition of $W(c^{f(1)}, l)$ the first half of each 2-block contains a monochromatic l-AP of 1-blocks. By similar reasoning, the first half of each 3-block contains a monochromatic l-AP of 2-blocks. This property holds at each level, with the first half of $\mathbf{N}(f(c + 1))$ containing a monochromatic l-AP of c-blocks. We consider only those integers that lie in l-APs at all c levels of blocks. We coordinatize each integer as

$$x = (x_1, \ldots, x_c),$$

with $1 \leq x_i \leq l$, where x_i is the position of x in the monochromatic l-AP of the i-block in which it occurs. All coordinatized integers have the same color, say α_1. Within each 1-block, the l integers

$$(1, x_2, \ldots, x_c), (2, x_2, \ldots, x_c), \ldots, (l, x_2, \ldots, x_c)$$

constitute a monochromatic l-AP. Therefore, the integer $(l + 1, x_2, \ldots, x_c)$ is a color other than α_1, say α_2. Furthermore, the factor 2 in the definition of $f(1)$ implies that $(l + 1, x_2, \ldots, x_c)$ occurs within the 1-block. Now we introduce the idea of focusing. Within a 2-block, the l integers

$$(l + 1, 1, x_3, \ldots, x_c), (l + 1, 2, x_3, \ldots, x_c), \ldots, (l + 1, l, x_3, \ldots, x_c)$$

are a monochromatic l-AP of color α_2. This forces $(l + 1, l + 1, x_3, \ldots, x_c)$ to be a color other than α_2. However, we can focus a second l-AP on this integer, namely,

$$(1, 1, x_3, \ldots, x_c), (2, 2, x_3, \ldots, x_c), \ldots, (l, l, x_3, \ldots, x_c).$$

Thus, $(l + 1, l + 1, x_3, \ldots, x_c)$ cannot be color α_1 or α_2; say it is color α_3. Figure 4.11 illustrates the two focused progressions, representing colors $\alpha_1, \alpha_2, \alpha_3$ by dots, circles, and an x, respectively. The dashes represent numbers with undetermined colors. Continuing the

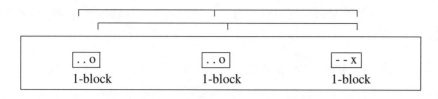

2-block

Figure 4.11. Focusing in a 2-block.

focusing process at each of the c levels, we conclude that $(l + 1, l + 1, \ldots, l + 1)$ can be none of the colors $\alpha_1, \ldots, \alpha_c$, a contradiction. Hence, there exists a monochromatic $(l + 1)$-AP. This completes the induction.

As an example of van der Waerden's theorem, there exists a number n such that no matter how $\mathbf{N}(n)$ is partitioned into two classes, one of the classes will contain a 3-AP. It turns out that the least such n is 9. That is, no matter how we partition $\mathbf{N}(9)$ into two classes, one of the two classes must contain a 3-AP; it is unavoidable. For example, if we have $\mathbf{N}(9) = \{1, 3, 4, 6, 8\} \cup \{2, 5, 7, 9\}$, then the first subset contains 4, 6, 8, which is a 3-AP.

But we can't take n to be 8 and always expect to get a 3-AP in one class. Here is a partition of $\mathbf{N}(8)$ into two classes neither of which contains a 3-AP:

$$\mathbf{N}(8) = \{1, 2, 3, 4, 5, 6, 7, 8\} = \{1, 3, 6, 8\} \cup \{2, 4, 5, 7\}.$$

This isn't the only partition that avoids a 3-AP in one class. We can also partition $\mathbf{N}(8)$ as

$$\{2, 4, 5, 7\} \cup \{1, 3, 6, 8\}.$$

It turns out that if we are looking for a 4-AP then we will have to partition $\mathbf{N}(35)$ into two classes in order to guarantee it. A general problem is the determination of $W(2, l)$ for various values of l. It has been solved only for $l = 1, 2, 3, 4,$ and 5. It is trivial that $W(2, 1) = 1$, because the class that contains the only element of $\mathbf{N}(1)$ will contain a 1-term arithmetic progression. It is also immediate that $W(2, 2) = 3$, because when three integers are partitioned into two classes, one of the classes must contain at least two integers and hence a 2-AP. We have indicated that $W(2, 3) = 9$. It is nontrivial to show that $W(2, 4) = 35$ and $W(2, 5) = 178$. The known values of $W(c, l)$ with $c \geq 2$ and $l \geq 3$ are $W(2, 3) = 9$, $W(2, 4) = 35$, $W(2, 5) = 178$, $W(3, 3) = 27$, and $W(4, 3) = 76$. As is suggested by the upper bound for $f(c)$ in our proof of van der Waerden's theorem, it is known that $W(c, l)$ grows very fast.

Van der Waerden's theorem is important in a branch of combinatorics called Ramsey theory. Ramsey theory is the part of combinatorics that studies the question, "What order exists in disorder?" Graphs and arithmetic sequences are natural settings the question, and there we find the seminal Ramsey theory results (Ramsey's theorem and van der Waerden's theorem). Van der Waerden's theorem says that, if to each positive integer we assign a color (say, green or red), then we will possess an infinite collection of pieces of information. The order in this collection is the existence of arbitrarily long monochromatic arithmetic progressions.

The definitive book on Ramsey theory, [22], explores several generalizations of van der Waerden's theorem.

4.15   Latin Squares and Projective Planes

A *latin square*[12] of order n is an $n \times n$ array in which all the numbers 1 through n appear in every row and in every column.

Two latin squares of order 3 are

$$
\begin{array}{ccc}
1 & 2 & 3 \\
2 & 3 & 1 \\
3 & 1 & 2
\end{array}
\quad \text{and} \quad
\begin{array}{ccc}
1 & 2 & 3 \\
3 & 1 & 2 \\
2 & 3 & 1.
\end{array}
$$

In each array, the numbers 1, 2, and 3 appear in every row and in every column.

If we superimpose them, every ordered pair of the numbers 1, 2, and 3 appears exactly once. Such latin squares are called *orthogonal*.

$$
\begin{array}{ccc}
11 & 22 & 33 \\
23 & 31 & 12 \\
32 & 13 & 21
\end{array}
$$

From the two orthogonal latin squares of order 3, we can construct a projective plane of order 3, as shown in Figure 2.5. Change the numbers in the latin squares from 1, 2, and 3 to 0, 1, and 2. A projective plane of order three has thirteen points and thirteen lines. The thirteen points are the ordered pairs

$$00, 01, 02, 10, 11, 12, 20, 21, 22,$$

together with the ideal points $\bar{0}, \bar{1}, \bar{2}$, and ∞. We arrange the nine non-ideal points in a 3×3 array, and the four ideal points around the array, as in Figure 2.5.

We need to define the thirteen lines. The first three lines are the horizontal lines

$$
\{00, 10, 20, \bar{0}\},
$$
$$
\{01, 11, 21, \bar{0}\},
$$
$$
\{02, 12, 22, \bar{0}\}.
$$

The next three lines are the vertical lines

$$
\{00, 01, 02, \infty\},
$$
$$
\{10, 11, 12, \infty\},
$$
$$
\{20, 21, 22, \infty\}.
$$

The next three lines go through constant values in the first latin square (together with $\bar{1}$):

$$
\{02, 10, 21, \bar{1}\},
$$
$$
\{01, 12, 20, \bar{1}\},
$$
$$
\{00, 11, 22, \bar{1}\}.
$$

---

[12]The term "latin square" derives from the use by Leonhard Euler (1707–1783) of latin letters instead of numbers in the arrays.

The next three lines go through constant values in the second latin square (together with $\bar{2}$):

$$\{02, 11, 20, \bar{2}\},$$
$$\{00, 12, 21, \bar{2}\},$$
$$\{01, 10, 22, \bar{2}\}.$$

Finally, we include the ideal line

$$\{\bar{0}, \bar{1}, \bar{2}, \infty\}.$$

It's easy to check that each line contains four points and each point is on four lines. Furthermore, it can be checked that two points determine a unique line and two lines intersect in exactly one point.

A set of *mutually orthogonal latin squares* (MOLS) of order n is a collection of latin squares in which each pair are orthogonal.

The maximum number of MOLS of order n is at most $n - 1$. To see this, suppose that we have n MOLS of order n. Relabel the numbers in each latin square (if necessary) so that the first row is $1, 2, \ldots, n$. Relabeling numbers doesn't change latinicity or orthogonality. Consider the $(2, 1)$ entry of each square. It cannot be 1 since there is a 1 above it. Hence, there are only $n - 1$ choices it and by the pigeonhole principle some two squares have the same entry, say m. But for these two squares the entry mm occurs twice in the superimposed squares, in the $(2, 1)$ position and the $(1, m)$ position, so they are aren't orthogonal. The contradiction implies that the maximum number of MOLS is at most $n - 1$.

Recall from Chapter 2 that a projective plane of order n is a collection of $n^2 + n + 1$ points and $n^2 + n + 1$ lines such that each line contains $n + 1$ points, each point lies on $n + 1$ lines, every two points determine a unique line, and every two lines intersect in exactly one point.

As in our example with $n = 3$, a projective plane of order n is equivalent to a set of $n - 1$ MOLS of order n.

Theorem.[13] *A projective plane of order n is equivalent to a set of n − 1 MOLS of order n.*

Starting with a collection of $n - 1$ MOLS of order n, the construction of a projective plane of order n works as in our example with $n = 3$. Let the MOLS be labeled $1, \ldots, n - 1$, with the entries in each latin square labeled $0, \ldots, n - 1$. Let the points of the projective plane be the ordered pairs (x, y), where $0 \leq x, y \leq n - 1$, together with the ideal points $\bar{0}$, $\bar{1}, \ldots, \overline{n - 1}, \infty$. This accounts for $n^2 + n + 1$ points. The lines are the n horizontal lines

$$\{(x, b): 0 \leq x \leq n - 1\} \cup \{\bar{0}\}, \quad 0 \leq b \leq n - 1,$$

the n vertical lines

$$\{(a, y): 0 \leq y \leq n - 1\} \cup \{\infty\}, \quad 0 \leq a \leq n - 1,$$

the ideal line

$$\{\bar{0}, \bar{1}, \ldots, \overline{n - 1}, \infty\},$$

---

[13] This theorem was first proved in 1938 by Raj Chandra Bose (1901–1987), who made fundamental contributions to the theory of algebraic designs and combinatorial codes.

and, for $1 \leq i \leq n - 1$, the collection of points whose latin square entries are constant in the ith square together with the corresponding ideal point:

$$\{(x, y): \text{entry } (x, y) \text{ of } i\text{th latin square equals } j\} \cup \{\overline{i}\}, \quad 0 \leq j \leq n - 1.$$

This accounts for $n + n + 1 + (n - 1)n = n^2 + n + 1$ lines.

It follows immediately from the definitions that each line contains $n + 1$ points and each point is on $n + 1$ lines.

Given any point P, the $n + 1$ lines containing P contain $(n + 1)n = n^2 + n$ points other than P. Because of latinicity and orthogonality, the points are distinct. Since these are all the points other than P, every point lies on a line with P. From the definitions, there is at most one line determined by any two given points. Hence, every two points determine a unique line.

Given any line l, there are $n + 1$ points on it and these points lie on $(n + 1)n$ other lines. Because two points determine exactly one line, these lines are distinct. Since these are all the lines other than l, every line intersects l. No two lines intersect in more than one point, for if two lines intersected in two points, then these two points would determine more than one line. Hence, every two lines intersect in exactly one point.

We have shown how $n - 1$ MOLS of order n yield a projective plane of order n. Now we go in the reverse direction. Suppose that we have a projective plane of order n. Choose a line, call it the x-axis, and label its $n + 1$ points as $0, 1, \ldots, n - 1, \overline{0}$. Choose another line through 0, call it the y-axis, and label its points $0, 1, \ldots, n - 1, \infty$, in such a way that the intersection of the two axes is labeled 0 on both axes. Call the line joining $\overline{0}$ and ∞ the ideal line. For every point P not on the ideal line, suppose that the line through P and ∞ (a "vertical line") intersects the x-axis at x, and the line through P and $\overline{0}$ (a "horizontal line") intersects the y-axis at y. Give P the coordinates (x, y). This also gives coordinates to the points on the axes. The ideal line has $n - 1$ points other than $\overline{0}$ and ∞. Each point will give rise to a latin square with coordinates (x, y), where $0 \leq x, y \leq n - 1$. Each of these points is on n lines other than the ideal line. Let the points on each line correspond to constant-value entries in the corresponding latin square. You may wish to confirm the latinicity and orthogonality properties of the resulting squares. We have shown that a projective plane of order n is equivalent to a set of $n - 1$ MOLS of order n.

It is always possible to construct a projective plane of order n, or the equivalent set of $n - 1$ MOLS of order n, from a field of order n. A field of order n exists if and only if n is a prime power (see Appendix A). However, there exist projective planes of prime power order that do not arise from fields. No one has found a finite projective plane that is not of prime power order or proven that none exists.

Here's the main idea of the construction of a set of $n - 1$ MOLS of order n from a field of order n. Let the nonzero field elements be $f_1, \ldots, f_{n-1}$. We define a latin square for each f_i, with $1 \leq i \leq n - 1$, by the rule that the (x, y) entry, where $0 \leq x, y \leq n - 1$, is

$$f_i f_x + f_y.$$

I leave it as an exercise to show that this produces mutually orthogonal latin squares. The two mutually orthogonal latin squares of order 3 shown at the beginning of this section arise from the field $\mathbf{Z}_3$.

For more on the connections between latin squares and finite geometries, see [29].

4.16   The Lemniscate Revisited

Recall the lemniscate graph from Chapter 1 (Figure 1.2), with parametric equations

$$x = \frac{\cos t}{1 + \sin^2 t}$$

$$y = \frac{\sin t \cos t}{1 + \sin^2 t}, \quad -\infty < t < \infty.$$

A moving point makes one lap around the graph as t goes from 0 to 2π. It repeats this figure-eight in periods of 2π. What distance does it travel? In other words, what is the arc length of the lemniscate graph?

To find the length of a curve, we integrate the differential of arc length,

$$\sqrt{\left(\frac{dx}{dt}\right)^2 + \left(\frac{dy}{dt}\right)^2} \, dt.$$

For the lemniscate, this is

$$\frac{dt}{\sqrt{1 + \sin^2 t}}.$$

The length L of the lemniscate curve is found by integrating the arc length over the interval of t values:

$$L = \int_0^{2\pi} \frac{dt}{\sqrt{1 + \sin^2 t}}.$$

Integrals of this type are known as *elliptic integrals* because they arise in connection with finding the arc length of an ellipse.

This isn't an easy integral to evaluate. If we use a computer algebra system to evaluate it numerically, we find an approximation to the length of the lemniscate curve:

$$L \doteq 5.244115108.$$

Carl Friedrich Gauss (1777–1855) discovered that this number is related to π. The circumference of a unit circle is

$$2\pi \doteq 6.283185307.$$

Hence, the ratio of the circumference of a unit circle to the length of the lemniscate curve is

$$\frac{2\pi}{L} \doteq 1.198140234.$$

This number may not look familiar, but Gauss recognized it as the *arithmetic-geometric mean* of 1 and $\sqrt{2}$.

Starting with two positive real numbers a and b, we define their arithmetic-geometric mean as follows. Set $a_0 = a$ and $b_0 = b$. Let a_1 be the arithmetic mean of a_0 and b_0, that is, $(a_0 + b_0)/2$; and let b_1 be the geometric mean of a_0 and b_0, that is, $b_1 = \sqrt{a_0 b_0}$. Repeat this process starting with the numbers a_1 and b_1. Thus $a_2 = (a_1 + b_1)/2$ and $b_2 = \sqrt{a_1 b_1}$. Continuing, we obtain sequences $\{a_n\}$ and $\{b_n\}$ that converge to the same limit, the arithmetic-geometric mean of a and b, denoted by $M(a, b)$. To see that the sequences converge, we use the fact that the arithmetic mean is always at least equal to the geometric mean (see page 55). Assuming that $b_0 \leq a_0$, we have

$$b_0 \leq b_1 \leq a_1 \leq a_0.$$

We see that $\{b_n\}$ is a nondecreasing sequence bounded above, and hence convergent. Similarly, $\{a_n\}$ is a nonincreasing sequence bounded below and hence convergent. Suppose that $\{a_n\}$ converges to L_1 and $\{b_n\}$ converges to L_2. Then $(L_1 + L_2)/2 = L_1$, and therefore $L_1 = L_2$, i.e., the two sequences converge to the same limit.

A calculation shows that the arithmetic-geometric mean of 1 and $\sqrt{2}$ is

$$M(1, \sqrt{2}) \doteq 1.198140234.$$

It can ne shown that $M(1, \sqrt{2})$ is a transcendental number (it isn't a zero of a polynomial with integer coefficients).

As Gauss did, we will demonstrate that

$$\frac{2\pi}{L} = M(1, \sqrt{2}).$$

The idea is to generalize the arc length integral. Often in mathematics, problems are more easily solved when they are generalized. Define

$$I(a, b) = \int_0^{\pi/2} \frac{dt}{\sqrt{a^2 \cos^2 t + b^2 \sin^2 t}}.$$

By symmetry, and since $1 + \sin^2 t = \cos^2 t + 2\sin^2 t$, the integral for the length of the lemniscate curve equals $4I(1, \sqrt{2})$.

The key step is to show that

$$I(a, b) = I\left(\frac{a+b}{2}, \sqrt{ab}\right) \quad (*).$$

Once this is accomplished, the rest of the demonstration will be easy. We can repeatedly use $(*)$ to obtain

$$I(a, b) = I(a_1, b_1) = I(a_2, b_2) = \cdots = I(M(a, b), M(a, b)).$$

Hence

$$I(a, b) = I(M(a, b), M(a, b))$$

$$= \int_0^{\pi/2} \frac{dt}{\sqrt{M(a, b)^2 \cos^2 t + M(a, b)^2 \sin^2 t}}$$

$$= \frac{1}{M(a, b)} \int_0^{\pi/2} dt$$

$$= \frac{\pi/2}{M(a, b)}.$$

Since for the lemniscate, $L = 4I(1, \sqrt{2})$, the relation $2\pi/L = M(1, \sqrt{2})$ follows immediately.

To finish the demonstration we must prove $(*)$. The change of variables

$$\sin t = \frac{2a \sin u}{a + b + (a - b) \sin^2 u},$$

where $0 \le t, u \le \pi/2$, yields

$$\cos t \, dt = \frac{2a \cos u (a + b + (b - a) \sin^2 u)}{(a + b + (a - b) \sin^2 u)^2} du,$$

and using some algebra and trigonometry we get

$$\frac{dt}{\sqrt{a^2 \cos^2 t + b^2 \sin^2 t}} = \frac{du}{\sqrt{\left(\frac{a+b}{2}\right)^2 - \left(\frac{a-b}{2}\right)^2 \sin^2 u}} = \frac{du}{\sqrt{a_1^2 \cos^2 u + b_1^2 \sin^2 u}}.$$

This proves $(*)$, so we can now state Gauss's discovery as a theorem.

Theorem. *Let L be the length of the lemniscate given by the parametric equations*

$$x = \frac{\cos t}{1 + \sin^2 t}$$

$$y = \frac{\sin t \cos t}{1 + \sin^2 t}, \quad 0 \le t \le 2\pi.$$

Then

$$\frac{2\pi}{L} = M(1, \sqrt{2}),$$

where $M(1, \sqrt{2})$ is the arithmetic-geometric mean of 1 and $\sqrt{2}$.

See [10] for a discussion of the relationship between elliptic integrals and elliptic curves.

5

Pleasing Proofs

Real mathematics ... must be justified as art if it can be justified at all.
—G. H. HARDY (1877–1947), *A Mathematician's Apology*[1]

In mathematics, assertions can be proved, which distinguishes mathematics from other disciplines. Mathematical knowledge is thus absolute and universal, independent of space and time. In this chapter, we present some proofs that are particularly memorable. Most are not well known and deserve to be better known.

5.1 The Pythagorean Theorem

The Pythagorean theorem states that given a right triangle, the area of a square formed on the hypotenuse is equal to the sum of the areas of the squares formed on the two legs.

There are many proofs of this important theorem. Figure 5.1 shows a tessellation proof. The plane is tessellated, or tiled, with copies of the square on the hypotenuse of the triangle (shaded in the figure), and also tessellated by copies of the squares on the two legs. This shows that the square on the hypotenuse can be divided into five pieces that can be reassembled to form the squares on the two legs. Two pieces make the smaller square and three pieces make the larger square.

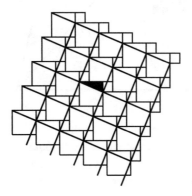

Figure 5.1. A tessellation proof of the Pythagorean theorem.

---

[1] Hardy, G.H., *A Mathematicians Apology*, pp. 139, Cambridge University Press, 1967. Reprinted with permission.

5.2 The Erdős–Mordell Inequality

In 1935 Paul Erdős conjectured a geometric inequality. Let ABC be a triangle and M be a point in the interior or on the boundary of ABC. Let the distances from M to the vertices A, B, C be x, y, z, respectively, and let the distances from M to the sides AB, BC, CA be c, a, b, respectively. Then

$$x + y + z \geq 2(p + q + r),$$

with equality occurring if and only if ABC is equilateral and M is its center. See the diagram below (the shading in the diagram will be explained in a moment).

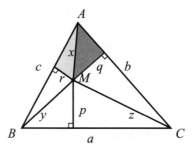

Soon after the conjecture was made, Louis Mordell and D. F. Barrow proved it, but the proof was complicated. Other proofs have been found from time to time, and in 2007 a simple proof was found by Claudi Alsina and Roger B. Nelsen [2]. This proof is so beautiful that it deserves to be admired.

The key idea is the inequality $ax \geq br + cq$, which we will now establish. Scale the triangle ABC by a factor of x, the light shaded triangle by a factor of b, and the dark shaded triangle by a factor of c; and put the triangles together as in the diagram below.

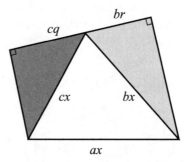

We claim that the figure is a trapezoid. To see that the base is really a straight line, note that the four angles at the vertex corresponding to A (looking at both diagrams) consist of two pairs of complementary angles, In a trapezoid, the side opposite the base is at least as long as the base, by the Pythagorean theorem. Therefore $ax \geq br + cq$. Equality occurs only when the trapezoid is a rectangle.

Similar arguments show that $by \geq cp + ar$ and $cz \geq aq + bp$. From the three inequalities,

$$x + y + z \geq \left(\frac{b}{c} + \frac{c}{b} \right) p + \left(\frac{a}{c} + \frac{c}{a} \right) q + \left(\frac{b}{a} + \frac{a}{b} \right) r.$$

By the arithmetic mean–geometric mean (AM–GM) inequality, the sums in parentheses are at least equal to 2. This proves the Erdős–Mordell inequality. Furthermore, equality occurs in the AM–GM inequality only if $a = b = c$, that is, only if ABC is an equilateral triangle. Given that ABC is an equilateral triangle and the trapezoid in the second diagram is a rectangle, the two right triangles in the second diagram are congruent and hence $q = r$. Similarly, $p = q$, and therefore M is the incenter (i.e., the center) of ABC.

In 2008 Victor Pambuccian [41] proved that in absolute geometry (Euclidean geometry without the parallel postulate), the Erdős–Mordell inequality is equivalent to the statement that the sum of the angles of a triangle is at most π.

5.3 Triangles with Given Area and Perimeter

If a triangle has perimeter 6, how large can its area be? With some thought, you can conclude that the unique triangle with perimeter 6 and maximum area is the equilateral triangle of side length 2. Its altitude is $\sqrt{3}$ and hence its area is $\sqrt{3}$. But let's change the question a little. What if we require a triangle of perimeter 6 and area equal to some specified number *less than* $\sqrt{3}$. It turns out that there are *infinitely many* triangles with perimeter 6 having the specified area.

In general, given positive real numbers A and P such that $A < \sqrt{3}P^2/36$, there are infinitely many triangles with area A and perimeter P. We'll prove this in a moment. First let's show that the area A cannot exceed $\sqrt{3}P^2/36$, where P is the perimeter. Recall Heron's formula for the area of a triangle (Chapter 3):

$$A = \sqrt{s(s-a)(s-b)(s-c)},$$

with a, b, c the sides lengths and $s = (a + b + c)/2$ the semiperimeter. By the arithmetic mean–geometric mean inequality (see Chapter 4), we have

$$A = \sqrt{s}\sqrt{(s-a)(s-b)(s-c)}$$

$$\leq \sqrt{s}\left(\frac{s-a+s-b+s-c}{3}\right)^{3/2}$$

$$= \sqrt{s}\left(\frac{s}{3}\right)^{3/2} = \frac{s^2}{3\sqrt{3}} = \frac{\sqrt{3}P^2}{36}.$$

This proves that the area of the triangle is at most $\sqrt{3}P^2/36$, and the maximum is attained only by an equilateral triangle with sides $P/3$.

Let's now prove that if A is any positive real number less than $\sqrt{3}P^2/36$, then there are infinitely many triangles with area A and perimeter P. Squaring both sides of Heron's formula, and using the fact that $P = 2s$, we obtain

$$16A^2 = P(P - 2a)(P - 2b)(P - 2c).$$

Since $P - 2c = 2a + 2b - P$, this becomes

$$16A^2 = P(P - 2a)(P - 2b)(2a + 2b - P).$$

Using the quadratic formula to solve for b, we have

$$b = \frac{P-a}{2} \pm \sqrt{a^2 - \frac{16A^2}{P(P-2a)}}.$$

One of the sign choices gives b and the other c.

The key observation is that this formula produce actual triangles if the quantity inside the square root is nonnegative, that is,

$$a^2 - \frac{16A^2}{P(P-2a)} \geq 0,$$

or

$$a^2 P(P-2a) \geq 16A^2.$$

We will show that there are infinitely many allowable choices for a, as long as $A < \sqrt{3}P^2/36$.

The function

$$f(a) = a^2 P(P-2a).$$

is a cubic polynomial with a double root at $a = 0$ and the third root at $a = P/2$. It is easy to show that f has a local maximum at $(P/3, P^4/27)$. If $16A^2 = P^4/27$, or equivalently $A = \sqrt{3}P^2/36$, then there is only one value of a for which $f(a) \geq 16A^2$, namely $a = P/3$, corresponding to the equilateral triangle of side $P/3$. However, if $A < \sqrt{3}P^2/36$, then there are infinitely many values of a (in an interval containing $P/3$) such that $f(a) \geq 16A^2$. Each such a yields a triangle with area A and perimeter P.

Notice that we didn't absolutely need the AM–GM inequality at the beginning of our analysis. Our argument boils down to Heron's formula, the quadratic formula, and simple properties of a cubic polynomial.

5.4   A Property of the Directrix of a Parabola

Given a parabola and a point in its "exterior," there are two tangent lines to the parabola passing through the point. The locus of points for which the two tangent lines are perpendicular is the parabola's directrix. We will prove these statements.

Without loss of generality, let the parabola be $y = kx^2$, and let $P = (x_0, y_0)$ be a point in the exterior of the parabola, that is, with $y_0 < kx_0^2$. A line through P tangent to the parabola at (a, ka^2) has equation

$$y_0 - ka^2 = m(x_0 - a),$$

where m is the slope. Since $m = 2ka$, we obtain

$$y_0 - \frac{m^2}{4k} = m\left(x_0 - \frac{m}{2k}\right),$$

or

$$m^2 - 4kx_0 m + 4ky_0 = 0.$$

The condition $y_0 < kx_0^2$ guarantees that the discriminant of the quadratic is positive; hence, there are two values of m corresponding to two tangent lines to the parabola through P.

These tangent lines are perpendicular if and only if the product of their slopes is -1. The product of the slopes is the constant coefficient of the quadratic, $4ky_0$. Thus, The two tangent lines passing through P are perpendicular if and only if $y_0 = -1/(4k)$, i.e., P is on the parabola's directrix.

We also give a geometric proof that every point on the directrix has the required property.

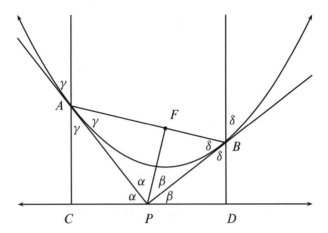

As in the diagram, let P be a point on the parabola's directrix. Draw tangent lines from P to the parabola at A and B, making angles α and β with the directrix. Let γ and δ be the complementary angles to α and β, respectively. Let vertical lines through A and B intersect the directrix at C and D, respectively. Hence, $\angle PAC = \gamma$ and $\angle PBD = \delta$. Construct line segments AF, BF, and PF, where F is the focus. A parabola has the property that a ray parallel to its axis of symmetry strikes the parabola and is reflected through the focus. This implies that $\angle PAF = \gamma$ and $\angle PBF = \delta$. Since $FA = AC$ (by definition of directrix), the triangles PAC and PAF are congruent. Hence $\angle FPA = \alpha$. By a similar argument, $\angle FPB = \beta$. Consequently, $2\alpha + 2\beta = 180°$ and thus $\angle APB$ is a right angle. (It also follows that $\gamma = \beta$, $\delta = \alpha$, and AFB is a straight line segment.)

It is easy to show that for any point exterior to the parabola and *not* on the directrix, the two tangent lines passing through it are not perpendicular. Hint: draw the intersection of one of the tangent lines with the directrix.

5.5 A Classic Integral

The classic integral formula

$$I = \int_{-\infty}^{\infty} e^{-x^2}\, dx = \sqrt{\pi}$$

can be proved by a clever trick that is well worth seeing.

The surprising technique is to evaluate a double integral:

$$I^2 = \int_{-\infty}^{\infty} e^{-x^2}\, dx \int_{-\infty}^{\infty} e^{-y^2}\, dy = \int_{-\infty}^{\infty}\!\int_{-\infty}^{\infty} e^{-x^2-y^2}\, dx\, dy.$$

We use polar coordinates r and θ, where $0 \le r < \infty$ and $0 \le \theta \le 2\pi$, with $x^2 + y^2 = r^2$ and $dx\,dy = r\,dr\,d\theta$. Then

$$
\begin{aligned}
I^2 &= \int_0^{2\pi} \int_0^\infty e^{-r^2} r\,dr\,d\theta \\
&= \int_0^{2\pi} d\theta \int_0^\infty r e^{-r^2}\,dr \\
&= 2\pi \left(-\frac{1}{2} e^{-r^2} \right) \Bigg]_0^\infty \\
&= 2\pi \cdot \frac{1}{2} \\
&= \pi.
\end{aligned}
$$

Therefore

$$
I = \sqrt{\pi}.
$$

The value of the integral can come as a surprise on first sight.

5.6   Integer Partitions

A *partition* of a positive integer n is a summation of positive integers equal to n. The order of the summands is unimportant. For example, the partitions of 5 are

$$5,\ 4+1,\ 3+2,\ 3+1+1,\ 2+2+1,\ 2+1+1+1,\ 1+1+1+1+1.$$

There is a rich literature on partitions of integers. Some good sources are [3] and [32]. One observation about partitions is that the number of partitions of n into odd parts (summands) is equal to the number of partitions of n into distinct parts. For instance, there are three partitions of 5 into odd parts:

$$5,\ 3+1+1,\ 1+1+1+1+1.$$

And there are three partitions of 5 into distinct parts:

$$5,\ 4+1,\ 3+2.$$

We will prove our claim for every positive integer n by describing a bijection (a one-to-one correspondence) between the set of partitions of n into odd parts and the set of partitions of n into distinct parts. Suppose that we have a partition of n into odd parts. If any two parts are equal, then replace them with their sum. Continue until we have distinct parts. In the other direction, starting with a partition of n into distinct parts, if a summand is even, replace it by two of its halves. Continue until we have only odd parts. This demonstrates the bijection.

The bijection associates partitions of 5 into odd parts and partitions of 5 into distinct parts as follows:

$$5 \leftrightarrow 5$$

$$3 + 1 + 1 \leftrightarrow 3 + 2$$

$$1 + 1 + 1 + 1 + 1 \leftrightarrow 4 + 1.$$

As an exercise, you could show the bijection for $n = 6$ (there are four partitions of each type).

Of course, we have to convince ourselves that the bijection is well defined. That is, if we start with a partition of n into odd parts, then there is only one way to do the combining and finish with a partition of n into distinct parts. And conversely, if we start with a partition of n into distinct parts, there is only one way to do the dividing and finish with a partition of n into distinct parts. And we must show that the two operations are inverses of each other. All this follows if we notice that combining two parts has no effect on the other parts, nor does dividing a part into two equal parts.

5.7 Integer Triangles

An *integer triangle* is a triangle with integer side lengths. Given a nonnegative integer n, let $t(n)$ be the number of incongruent integer triangles with perimeter n. For example, $t(10) = 2$, since there are two integer triangles with perimeter 10, namely, $(3, 3, 4)$ and $(2, 4, 4)$. We write a triangle as an ordered triple (a, b, c), with $a \leq b \leq c$. Here are the first few values of the sequence $\{t(n)\}$, known as *Alcuin's sequence*.

n	0	1	2	3	4	5	6	7	8	9	10
$t(n)$	0	0	0	1	0	1	1	2	1	3	2

What is a formula for $t(n)$? I will give the simplest derivation that I know of. It relates integer triangles to partitions of an integer into three parts. The key observation is that

$$t(2n) = u(n), \quad n \geq 0,$$

where $u(n)$ is the number of ways to write n as a sum of three positive integers (order unimportant). For example, $t(10) = u(5) = 2$, as there are two partitions of 5 into three parts, namely,

$$2 + 2 + 1 \quad \text{and} \quad 3 + 1 + 1.$$

The function $u(n)$ is often denoted by $p(n, 3)$ or $p_3(n)$.

We see that $t(2n) = u(n)$ for all $n \geq 0$ by the bijection

$$(a, b, c) \leftrightarrow \{n - a, n - b, n - c\},$$

where (a, b, c) represents the sides of an integer triangle of perimeter $2n$ and $\{n - a, n - b, n - c\}$ represents the summands in a partition of n into three parts. Notice that a, b, and c satisfy the triangle inequality if and only if each quantity is less than n, and this is true if and only if $n - a$, $n - b$, and $n - c$ are positive.

Next, we will show that $\{u(n)\}$ satisfies a simple recurrence relation. Consider three cases according to the size of the last summand in a partition of $n + 6$ into three parts, where the parts are written in non-increasing order. If the last summand is 3 or greater, then subtract 2 from each summand to obtain a partition of n into three parts. If the last summand is a 2, then subtract 2 from each part to obtain a partition of n into one or two parts. If the last summand is a 1, then subtract 1 from each part to obtain a partition of $n + 3$ into one or two parts. Given any positive integer m, the number of partitions of m into one part is 1, and the number of partitions of m into two parts is $\lfloor m/2 \rfloor$, where $\lfloor x \rfloor$, the *floor* of x, is the greatest integer less than or equal to x. Hence

$$u(n + 6) = u(n) + 1 + \left\lfloor \frac{n}{2} \right\rfloor + 1 + \left\lfloor \frac{n + 3}{2} \right\rfloor, \quad n \geq 0.$$

We can simplify by examining the cases n even and n odd, to obtain the recurrence relation

$$u(n + 6) = u(n) + n + 3, \quad n \geq 0.$$

Remembering that $u(n) = t(2n)$, we have the initial values

n	0	1	2	3	4	5
$u(n)$	0	0	0	1	1	2

Using the recurrence relation and the initial values, we can generate data and guess the formula

$$u(n) = \left[\frac{n^2}{12} \right], \quad n \geq 0,$$

where $[x]$ denotes the nearest integer to x. Because $n^2/12$ is never a half-integer, the formula is well-defined. The formula has the correct six initial values. All that remains to show is that it satisfies the recurrence relation for $\{u(n)\}$. This is simple:

$$\left[\frac{(n + 6)^2}{12} \right] = \left[\frac{n^2 + 12n + 36}{12} \right] = \left[\frac{n^2}{12} \right] + n + 3.$$

By our analysis so far, we have found a formula for $t(n)$ with n even:

$$t(n) = u(n/2) = [n^2/48].$$

Taking another look at the data, we conjecture that

$$t(n) = t(n + 3), \quad \text{if } n \text{ is odd}.$$

This gives us a way to calculate $t(n)$ for n odd. You can establish the relation by showing the correspondence

$$(a, b, c) \leftrightarrow (a + 1, b + 1, c + 1),$$

where (a, b, c) represents an integer triangle with odd perimeter n.

Putting the pieces together, we obtain a formula for the number of incongruent integer triangles with perimeter n:

$$t(n) = \begin{cases} \left[\dfrac{n^2}{48} \right] & \text{if } n \text{ is even}, \\ \left[\dfrac{(n+3)^2}{48} \right] & \text{if } n \text{ is odd}, \end{cases} \quad n \geq 0.$$

Alcuin's sequence is named in honor of Alcuin of York (732–804). Alcuin is credited with writing a book called *Propositions of Alcuin, A Teacher of Emperor Charlemagne, for Sharpening Youths.* The book contains 53 mathematical word problems that can be solved by simple arithmetic, algebra, or geometry. The twelfth problem is an allocation problem.

Problem 12: A father and his three sons. *A father, when dying, gave to his sons* 30 *glass flasks, of which* 10 *were full of oil,* 10 *were half full, and* 10 *were empty. Divide the oil and the flasks so that the three sons receive the same number of flasks and the same amount of oil.*

Each son receives 10 flasks. There are five solutions, which can be listed by the number of full flasks that go to each son: $\{5, 5, 0\}$, $\{5, 4, 1\}$, $\{5, 3, 2\}$, $\{4, 4, 2\}$, and $\{4, 3, 3\}$. We don't count permutations of solutions as distinct. Each son receives an equal number of full flasks and empty flasks, and a number of half-empty flasks to make the total 10.

The number of ways to allocate n full flasks of oil, n half-full flasks, and n empty flasks to three persons so that each person receives the same number of flasks and the same amount of oil is $t(n + 3)$. This is why $\{t(n)\}$ is called Alcuin's sequence.

Alcuin's sequence has many wonderful properties. For example, it is a *zigzag sequence* (its values alternately rise and fall starting with $n = 6$). It can be extended to negative values of n using the same formula that we found, and the doubly-infinite sequence is *palindromic*, i.e., it satisfies the same recurrence relation going forward as backward. The order nine linear recurrence relation with constant coefficients is

$$t(n) = t(n - 2) + t(n - 3) + t(n - 4)$$
$$- t(n - 5) - t(n - 6) - t(n - 7) + t(n - 9), \quad n \geq 9.$$

We can see the form of this recurrence relation in the expansion of the denominator of the rational generating function for Alcuin's sequence:

$$\frac{x^3}{(1 - x^2)(1 - x^3)(1 - x^4)} = \frac{x^3}{1 - x^2 - x^3 - x^4 + x^5 + x^6 + x^7 - x^9}.$$

The *period* of a sequence $\{a(n)\}$ is the least positive integer k such that $a(n + k) = a(n)$ for all n. Given $m \geq 2$, the period of the sequence $\{t(n) \mod m\}$ is $12m$. For example, Alcuin's sequence modulo 2 begins

$$0, 0, 0, 1, 0, 1, 1, 0, 1, 1, 0, 0, 1, 1, 0, 1, 1, 0, 1, 0, 0, 0, 0, 0,$$

and the pattern repeats with a period of 24. This is an intriguing result because the corresponding problem for the Fibonacci sequence remains unsolved, except for moduli that are powers of 2 or 5.

Let's prove the claim. We first show that the sequence $\{t(n) \mod m\}$ repeats in a cycle of length $12m$. It follows that its period is a divisor of $12m$. For even values of the argument, we have

$$t(2n + 12m) = [(2n + 12m)^2/48] = [(2n)^2/48] + m + 3m^2 \equiv t(2n) \pmod{m}.$$

For n odd we have

$$t(n + 12m) = t(n + 3 + 12m) \equiv t(n + 3) = t(n) \pmod{m}.$$

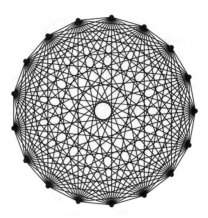

Figure 5.2. The complete graph of order 17.

We next show that the period of $\{t(n) \mod m\}$ is not a proper divisor of $12m$. Since the period is a divisor of $12m$, it is of the form $12r$, for $r \leq m$, or the sequence repeats in a cycle of length $6m$ or $4m$. If the period has the form $12r$, then m divides $t(12r) = 3r^2$ (since $t(0) = 0$) and m divides $t(12r + 2) = 3r^2 + r$ (since $t(2) = 0$). Hence, m divides the difference, which is r, and $m \leq r$.

We will show that the sequence $\{t(n) \mod m\}$ does not repeat in a cycle of length $6m$. We have already covered the case $m = 2s$. Let $m = 4s + 1$. Then $6m = 24s + 6$ and $t(6m) = 12s^2 + 6s + 1 = 3s(4s + 1) + 3s + 1$, and we see that m does not divide $t(6m)$. The case $m = 4s + 3$ is similar.

Finally, we will show that the sequence $\{t(n) \mod m\}$ does not repeat in a cycle of length $4m$. We have already covered the case $m = 3s$. Let $m = 3s + 1$. Then $4m = 12s + 4$ and $t(4m) = s(3s + 1) + s$, and again we see that m does not divide $t(4m)$. The case $m = 3s + 2$ is similar.

We conclude that the period of $\{t(n) \mod m\}$ is $12m$.

It can be shown (see [8]) that the sequence $\{t(n) \mod m\}$ takes every value modulo m if and only if m is one of

$$7, \ 10, \ 19, \ 2^j, \ 3^j, \ 5^j, \ 11^j, \ 13^j, \ 41^j, \ 2 \cdot 3^j, \ 5 \cdot 3^j, \quad \text{for } j \geq 1.$$

5.8   Triangle Destruction

The *complete graph* of order n consists of n *vertices* and all possible *edges* between pairs of vertices. The edges can be drawn straight or curved and may cross. Recall Figure 2.6, the complete graph on 17 points. Figure 5.2 shows this graph without the two-coloring of edges.

Consider the game called *Triangle Destruction*, played on the graph. Two players, Oh and Ex, alternately remove edges. Oh moves first, removing an edge. Then Ex moves, removing any other edge. Then Oh removes an edge, and so on. The first player to eliminate the last triangle of the graph is the winner. Given best possible play by both players, who should win Triangle Destruction?

The key is to think of what the graph looks like immediately before the second-to-last move. The player whose turn it is cannot win, but no matter what move he or she makes, the other player can then win immediately. The graph before the second-to-last move must contain two triangles, for if it contained no triangles then the game would already be over, and if it contained only one triangle then the player whose turn it was would destroy that triangle and win.

There are three cases.

Case 1: The graph contains two disjoint triangles.

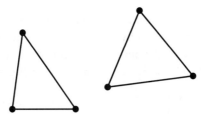

In this case, the graph contains no other edges, or else the player to move would delete one of the extraneous edges and not lose on the next turn.

Case 2: The graph contains two triangles with a vertex in common.

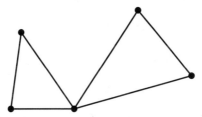

This is essentially the same as Case 1. The graph contains no other edges, or else the player to move would delete one of the extraneous edges and not lose on the next turn.

Case 3: The graph contains two triangles that share an edge.

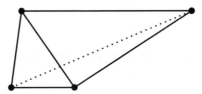

In this case, we have a collection of four vertices and five edges. The graph contains no edges outside the complete graph on these four vertices, or else the player on the move would delete one of the extraneous edges and not lose on the next turn. The sixth edge (the dotted line), making a complete graph, would also have to be in the graph, or else the player to move would remove the edge adjacent to both triangles and destroy them both.

In all cases, the graph at this stage has six edges, an even number. Since we start with a graph with $\binom{17}{2} = 136$ edges, an even number of turns must have taken place. Therefore, it must be Oh's turn. Hence Ex will win.

Triangle Destruction can be played on the complete graph of order n. By the same reasoning as in the $n = 17$ game, Ex wins if $n \equiv 0, 1 \pmod 4$, and Oh wins if $n \equiv 2, 3 \pmod 4$.

5.9   Squares in Arithmetic Progression

The numbers 1, 25, and 49 are three perfect squares in arithmetic progression (since $25 - 1 = 49 - 25$). The same is true of the trio 4, 100, and 196. There are infinitely many triples of distinct perfect squares of positive integers in arithmetic progression, as evidenced by

$$(a^2 - 2ab - b^2)^2, \quad (a^2 + b^2)^2, \quad (b^2 - 2ab - a^2)^2,$$

where a and b are positive integers with $a > b$. Clearly, these are squares. We can check that they are in arithmetic progression by expanding the expressions and subtracting the first from the second and the second from the third. Both differences are $4ab(a^2 - b^2)$.

For example, with $a = 7$ and $b = 5$, we obtain the arithmetic progression

$$46^2 = 2116, \quad 74^2 = 5476, \quad 94^2 = 8836,$$

with common difference 3360.

How do we find such a solution? We use a number theory technique, due to Diophantus, in which we can find rational points on a curve given a rational base point. Let the three numbers be x^2, y^2, and z^2, with $z^2 - y^2 = y^2 - x^2$, so that

$$x^2 + z^2 = 2y^2.$$

Let $x' = x/y$ and $z' = z/y$, so that

$$x'^2 + z'^2 = 2.$$

We have a solution $(x', z') = (1, 1)$. Suppose that there is another solution (m, n), where m and n are rational numbers. Then the slope of the line from $(1, 1)$ to (m, n) is a rational number, say t. Thus $t = (n - 1)/(m - 1)$, and we have $n = t(m - 1) + 1$. Hence

$$m^2 + [t(m - 1) + 1]^2 = 2,$$

and it follows that

$$(1 + t^2)m^2 + (2t - 2t^2)m + (t^2 - 2t - 1) = 0.$$

By the quadratic formula,

$$m = \frac{2t^2 - 2t \pm 2(1 + t)}{2(1 + t^2)}.$$

The positive sign gives us the solution we already know, $(1, 1)$, so the negative sign applies:

$$m = \frac{t^2 - 2t - 1}{t^2 + 1}, \quad n = \frac{-t^2 - 2t + 1}{t^2 + 1}.$$

Letting $t = a/b$, where a and b are integers ($b \neq 0$), we obtain

$$m = \frac{a^2 - 2ab - b^2}{a^2 + b^2}, \quad n = \frac{b^2 - 2ab - a^2}{a^2 + b^2}.$$

Multiplying these values by $a^2 + b^2$ results in the solution.

While there are infinitely many sets of three squares in arithmetic progression, there are no sets of four squares in arithmetic progression. This was conjectured by Pierre de Fermat (1601–1665) and proved by Leonhard Euler (1707–1783).

An application of Diophantus' method is the parameterization of all rational points on the unit circle $x^2 + y^2 = 1$. Taking $(0, -1)$ as the base point, we can show that all rational points on the unit circle, except the base point, are given by

$$x = \frac{2t}{1 + t^2}, \quad y = \frac{1 - t^2}{1 + t^2}, \quad t \in \mathbf{Q}.$$

If we allow $t = \infty$, then we get the point $(0, -1)$.

Similarly, we can parameterize all rational points on the unit sphere $x^2 + y^2 + z^2 = 1$. With $(0, 0, -1)$ as the base point, we obtain

$$x = \frac{2s}{1 + s^2 + t^2}, \quad y = \frac{2t}{1 + s^2 + t^2}, \quad z = \frac{1 - s^2 - t^2}{1 + s^2 + t^2}, \quad s, t \in \mathbf{Q}.$$

If $s = t = \infty$, then we get the point $(0, 0, -1)$.

If we allow s and t to be real numbers, then the parameterization is essentially the stereographic projection of the plane onto the unit sphere given in Riemann Sphere in Chapter 2.

5.10   Random Hemispheres

Let n hemispheres of a fixed sphere be selected at random. The probability that the sphere is covered by the hemispheres is $1 - 2^{-n}(n^2 - n + 2)$. Here is a proof: Each hemisphere is bounded by a great circle. The n great circles divide the surface of the sphere into $n^2 - n + 2$ regions (which can be proved by mathematical induction). Given a great circle, there are two hemispheres that have it as a boundary. Hence, the probability that a given region within the great circle is covered by one of the two hemispheres is $1/2$. The probability that the region is covered by none of the n hemispheres is 2^{-n}. Since there are $n^2 - n + 2$ regions, the probability that at least one of them is covered by no hemisphere is $2^{-n}(n^2 - n + 2)$. The probability that the entire sphere is covered is the complementary probability.

The event that a sphere is not covered by n hemispheres is the same as the event that the n centers of the hemispheres are contained in a single hemisphere (why?). So we have shown that the probability that n random points on the surface of a sphere are contained in a hemisphere is $2^{-n}(n^2 - n + 2)$.

As a generalization, suppose that n spherical caps are selected at random on a sphere. Let the subtended central angle of each cap be 2α, where $0 < \alpha < \pi$. If $\alpha = \pi/2$, then each cap is a hemisphere. Let $f(n)$ be the probability that the surface of the sphere is covered by the n spherical caps. No exact formula for $f(n)$ is known, but an asymptotic result is

$$\lim_{n \to \infty} \frac{\log(1 - f(n))}{n} = \log\left(1 - \sin^2 \frac{\alpha}{2}\right),$$

where log is the natural logarithm.

5.11   Odd Binomial Coefficients

A positive integer n *dominates* a positive integer k if the powers of 2 expansion of n contains all the terms in the powers of 2 expansion of k. For example, 45 dominates 9, since

$45 = 32 + 8 + 4 + 1$ and $9 = 8 + 1$. The binomial coefficient $\binom{n}{k}$ is odd if and only if n dominates k.

One way to prove this is by showing that the exact power of 2 that divides $\binom{n}{k}$ is equal to the number of carries when k and $n - k$ are added in base 2. We claim that

$$(x + 1)^{2^j} \equiv x^{2^j} + 1 \pmod{2}, \quad j \geq 0.$$

The reason is that in the binomial expansion of the expression on the left, the binomial coefficients $\binom{2^j}{m}$, for $1 < m < 2^j$, are all even. This is because

$$\binom{2^j}{m} = \frac{2^j \binom{2^j - 1}{m - 1}}{m},$$

and m divides the product in the numerator on the right yet $m < 2^j$.

Consider a specific value of n, such as 45. We have

$$\begin{aligned}
(x + 1)^{45} &= (x + 1)^{32 + 8 + 4 + 1} \\
&= (x + 1)^{32}(1 + x)^8(1 + x)^4(1 + x)^1 \\
&\equiv (1 + x^{32})(1 + x^8)(1 + x^4)(1 + x^1) \pmod{2}.
\end{aligned}$$

Modulo 2, the only binomial coefficients that show up on the left are the odd ones, but on the right the only binomial coefficients $\binom{45}{m}$ that show up are those for which 45 dominates m. This proves the result in the case $n = 45$, and the general result is proved in the same way.

An immediate consequence of our result is that the number of odd entries in the nth row of Pascal's triangle is 2^α, where α is the number of 1s in the binary representation of n.

By the way, if in Pascal's triangle you replace the even numbers by 0s and the odd numbers by 1s, you will get a pattern that looks like Sierpiński's triangle of Chapter 2.

5.12 Frobenius' Postage Stamp Problem

Suppose that postage stamps come in denominations a and b, where a and b are positive integers greater than 1 with greatest common divisor 1, and we have an unlimited supply of stamps.

The problem is to find those positive integer amounts that cannot be made using these stamps.[2]

(a) The number of amounts that cannot be made is finite.

(b) The largest amount that cannot be made is $ab - a - b$.

(c) The number of amounts that cannot be made is $(ab - a - b + 1)/2$.

For example, if $a = 3$ and $b = 5$, then we can make all amounts except 1, 2, 4, and 7.

---

[2]The creator of this problem, Ferdinand Georg Frobenius (1849–1917), made contributions in the areas of differential equations and group theory.

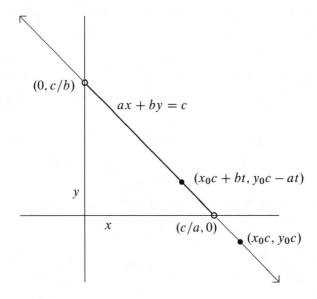

Figure 5.3. The line $ax + by = c$.

If $\gcd(a, b) \neq 1$, we would have infinitely many amounts that cannot be made. For example, if $a = 2$ and $b = 4$, then no odd amount can be made.

(a) We will show that if $c > ab$, then the amount c can be made with a positive number of a stamps and b stamps. It follows that the number of positive integer amounts that cannot be made is finite.

A well-known theorem of number theory is if a and b are relatively prime positive integers, i.e., $\gcd(a, b) = 1$, then there exist integers x_0 and y_0 such that $ax_0 + by_0 = 1$. Assume that $c > ab$. We want to show that there exist positive integers x and y such that $ax + by = c$. This is equivalent to showing that the line $ax + by = c$ contains a point with integer coordinates in the first quadrant of the plane. See Figure 5.3.

For any integer t we have

$$a(x_0 c + bt) + b(y_0 c - at) = c.$$

So integer solutions exist along the line shown at intervals of step length $\sqrt{a^2 + b^2}$. The result will follow if we show that the length of the line segment in the first quadrant is greater than the step length. Thus

$$c > ab \implies \frac{c}{ab}\sqrt{a^2 + b^2} > \sqrt{a^2 + b^2} \implies \sqrt{\left(\frac{c}{a}\right)^2 + \left(\frac{c}{b}\right)^2} > \sqrt{a^2 + b^2}.$$

Therefore, some integer solution lies in the first quadrant, and we are done.

(b) A generating function gives the result. Without loss of generality, assume that $a < b$. The generating function for the nonnegative integer amounts that can be made using no b stamps (and any number of a stamps) is

$$1 + x^a + x^{2a} + \cdots = \frac{1}{1 - x^a}.$$

The generating function for the amounts that can be made with one b stamp is

$$x^b + x^{a+b} + x^{2a+b} + \cdots = \frac{x^b}{1 - x^a}.$$

The generating function for the amounts that can be made with two b stamps is

$$x^{2b} + x^{a+2b} + x^{2a+2b} + \cdots = \frac{x^{2b}}{1 - x^a}.$$

... The generating function for the amounts that can be made with $a - 1$ of the b stamps is

$$x^{(a-1)b} + x^{a+(a-1)b} + x^{2a+(a-1)b} + \cdots = \frac{x^{(a-1)b}}{1 - x^a}.$$

We can stop here, because the amounts that can be made with a or more of the b stamps have already been counted (using b or more of the a stamps). Hence, the generating function for all nonnegative integer amounts that can be made is

$$\frac{1}{1 - x^a} \left(1 + x^b + x^{2b} + \cdots + x^{(a-1)b} \right) = \frac{1 - x^{ab}}{(1 - x^a)(1 - x^b)}.$$

The generating function for all nonnegative integer amounts is

$$1 + x + x^2 + \cdots = \frac{1}{1 - x}.$$

Therefore, the generating function for all amounts that cannot be made is

$$\frac{1}{1 - x} - \frac{1 - x^{ab}}{(1 - x^a)(1 - x^b)} = \frac{(1 - x^a)(1 - x^b) - (1 - x)(1 - x^{ab})}{(1 - x)(1 - x^a)(1 - x^b)}.$$

We know from (a) that this rational function is a polynomial. The highest amount that cannot be made using a and b stamps is the degree of this polynomial, i.e., the degree of the numerator minus the degree of the denominator:

$$(ab + 1) - (a + b + 1) = ab - a - b.$$

(c) The number of amounts that cannot be made is the number of nonzero terms in the polynomial. We can obtain this by evaluating it at $x = 1$, or, equivalently, by calculating the limit of the generating function as $x \to 1$. Since both numerator and denominator tend to 0, we use l'Hôpital's rule, and we must do so three times. The third-order derivative of the numerator, letting $x = 1$, is

$$-3ab(ab - a - b + 1).$$

The third order derivative of the denominator, letting $x = 1$, is

$$-6ab.$$

The quotient is

$$\frac{ab - a - b + 1}{2},$$

so this is the number of positive integer amounts that cannot be made using a and b stamps.

The generalization of Frobenius' problem to three stamp denominations is unsolved. See [40] for an algorithmic approach to the problem and a survey of results.

5.13   Perrin's Sequence

Let $\{a_n\}$ be the nth term of *Perrin's sequence* (named after the French mathematician R. Perrin), defined by

$$a_0 = 3, \ a_1 = 0, \ a_2 = 2,$$

$$a_n = a_{n-2} + a_{n-3}, \ n \geq 3.$$

If p is prime, then $p \mid a_p$ (p divides a_p). See [34] for three pleasing proofs of this Perrin property.

We give a simple counting argument based on the fact that a_n is the number of maximal independent subsets of $\{1, 2, 3, \ldots, n\}$, as observed by Zoltán Füredi in 1987.

Denote by b_n the number of maximal independent subsets of $\{1, 2, 3, \ldots, n\}$, where $n \geq 2$. The word *independent* means that the subset contains no two consecutive numbers (where 1 and n are regarded as consecutive). The word *maximal* means that no more elements can be included in the subset. An example of a maximal independent subset of $\{1, 2, 3, 4, 5, 6, 7\}$ is $\{1, 4, 6\}$.

We will prove that $a_n = b_n$, for $n \geq 2$, and show that $p \mid a_p$, for p prime.

However, $n \mid a_n$ does not imply that n is prime. The smallest counterexample is $n = 271441 = 521^2$.

We can easily check that $b_2 = 2$, $b_3 = 3$, and $b_4 = 2$, agreeing with Perrin's sequence.

Suppose that the largest element in a maximal independent subset of $\{1, 2, 3, \ldots, n\}$ is k and the second-largest element is j. Then clearly $k = j + 2$ or $k = j + 3$. Eliminating k and the numbers between k and j results in a maximal independent subset of a set of size $n - 2$ (if $k = j + 2$) or size $n - 3$ (if $k = j + 3$). This transformation, and its inverse, show that $b_n = b_{n-2} + b_{n-3}$, for $n \geq 3$. Therefore $a_n = b_n$, for $n \geq 2$.

Now we can give a combinatorial proof that a prime p divides the pth Perrin number. Let p be a prime. Consider the map, say f, that acts on maximal independent subsets of $\{1, \ldots, p\}$ by rotating them one step forward: $f(k) = k + 1$ if $k < p$, and $f(p) = 1$. We claim that f has no fixed points; that is, there exists no maximal independent subset S of $\{1, \ldots, p\}$ such that f applied to S is S itself. The reason is that if there were such an S, then, given any element $s \in S$, by definition of f we would have $s + 1 \in S$. But this implies that S contains all the elements of $\{1, \ldots, p\}$ and hence could not be independent. Furthermore, we claim that each independent subset S of $\{1, \ldots, p\}$ has order p under f. This means that f must be applied p times to S in order to obtain S again. I leave this as an exercise (it uses the fact that p is a prime). These observations imply that the collection of maximal independent subsets of $\{1, \ldots, p\}$ is partitioned into sub-collections of size p, which means that p divides the number of maximal independent subsets of $\{1, \ldots, p\}$, that is, $p \mid a_p$.

5.14   On the Number of Partial Orders

A *partial order* on a set is a binary relation that is transitive, reflexive, and anti-symmetric (see Appendix A). For example, the relation "a divides b" is a partial order on the set $\{1, 2, 3, 4, 5, 6\}$. Let $p(n)$ be the number of partial orders on the set $\{1, \ldots, n\}$. No formula for $p(n)$ is known. The known values of $p(n)$, at the time of this writing, are given in the table. See entry A001035 in the Online Encyclopedia of Integer Sequences.

n	$p(n)$
1	1
2	3
3	19
4	219
5	4231
6	130023
7	6129859
8	431723379
9	44511042511
10	6611065248783
11	1396281677105899
12	414864951055853499
13	171850728381587059351
14	98484324257128207032183
15	77567171020440688353049939
16	83480529785490157813844256579
17	122152541250295322862941281269151
18	241939392597201176602897820148085023

The units digits appear to be periodic with period 4. The repeating block is 1, 3, 9, 9. Can we prove this?

When faced with a challenging mathematical question, it's often a good idea to generalize, which may give a better understanding. Trials with other moduli suggest that if the modulus m is a prime number, then the sequence is periodic with period $m - 1$. If the modulus m is a prime power, the sequence appears to be periodic with period $\phi(m)$, where ϕ is Euler's ϕ-function. (This function counts the number of positive integers less than m that have no common factor with m.) For any modulus m, the sequence appears to be periodic with period equal to the least common multiple (lcm) of the constituent periods. For example, with $m = 12$, the period appears to be $\mathrm{lcm}(\phi(4), \phi(3)) = \mathrm{lcm}(2, 2) = 2$.

Let's prove our conjecture when m is prime. This result was originally proved by Z. I. Borevich in 1984, but we will give a beautiful 2010 proof (unpublished) by Aaron Meyerowitz. It is similar to the proof of a property of Perrin's sequence given in the previous section.

We will show that

$$p(n + m - 1) \equiv p(n) \pmod{m}, \quad n \geq 1,$$

where m is prime. It's easier to follow the proof with specific numbers, so let's take $m = 5$ and $n = 3$. We want to show that

$$p(7) \equiv p(3) \pmod 5.$$

Remember that $p(3)$ is the number of partial orders on the set $\{1, 2, 3\}$. The key idea is to replace the 3 in this set by five clones, say $3a$, $3b$, $3c$, $3d$, and $3e$. Now we have a set of seven elements: $\{1, 2, 3a, 3b, 3c, 3d, 3e\}$. We will consider partial orders on this set, which is equivalent to the set $\{1, 2, 3, 4, 5, 6, 7\}$. Let σ be the permutation of the set $\{1, 2, 3a, 3b, 3c, 3d, 3e\}$ that fixes 1 and 2, and cycles the clones:

$$\sigma = (1)(2)(3a, 3b, 3c, 3d, 3e).$$

It induces a permutation on the collection of all partial orders on the set $\{1, 2, 3a, 3b, 3c, 3d, 3e\}$. Since σ is a 5-cycle, and 5 is a prime, its orbits have sizes 1 or 5. We are counting modulo 5, so the orbits of size 5 may be ignored. When can σ result in an orbit of size 1 (a fixed point)? If a partial order is fixed, then there can be no relations among the clones (other than the reflexive relations), or else a clone with no predecessor would be mapped to another clone that has a predecessor. Furthermore, each clone must act exactly like the element 3 in the set $\{1, 2, 3\}$. Hence, the orbits of size 1 are in bijective correspondence with the partial orders on $\{1, 2, 3\}$. This proves the congruence, and the general case where m is a prime is proved in the same manner.

Can this argument can be boosted up (say, by defining clones of clones) to account for the case where m is a prime power? For our purposes, we need the case only when m is prime. Applying our result with $m = 2$ and $m = 5$, we find that $p(n+1) - p(n)$ is divisible by 2 for all n (thus, all $p(n)$ are odd), and $p(n+4) - p(n)$ is divisible by 5 for all n. It follows that $p(n + 4) - p(n)$ is divisible by 10 for all n, and hence the block 1, 3, 9, 9 repeats forever. It is gratifying that we can prove such a thing about a sequence for which we can't compute many values.

5.15   Perfect Error-Correcting Codes

A *binary code* is a set of binary vectors of some fixed length. For example, the set

$$\{(0, 0, 1),\ (1, 1, 0),\ (1, 1, 1)\}$$

constitutes a code. The elements of a code are called *codewords*.

If we have information to send over a noisy channel, one in which errors occur, we can first encode the information with codewords. The *distance* between two binary strings is the number of coordinates in which they differ. For example, the distance between $(0, 0, 1)$ and $(1, 1, 1)$ is 2, since their first and second coordinates differ. The *distance* of a code is the minimum distance between any two codewords. The code given above has distance 1.

If a code has distance $2e + 1$, then there is a method for correcting e or fewer errors (alterations of bits). If e or fewer errors occur in the transmission of a codeword, we assume that the intended codeword is the one within distance e of the received binary vector.

We think of each codeword as surrounded by a sphere of radius e, consisting of all binary vectors whose distance from the center is at most e. All such vectors are decoded to the codeword at the center. If the vectors have length n, then the number of binary vectors in a sphere of radius e is

$$\sum_{i=0}^{e} \binom{n}{i},$$

since, for $0 \leq i \leq e$, there are $\binom{n}{i}$ choices for which i coordinates of a binary vector disagree with the codeword.

A code is capable of correcting e errors if and only if the spheres of radius e around the codewords are disjoint. Since there are 2^n binary strings of length n, in a code with w codewords it is necessary that

$$w \sum_{i=0}^{e} \binom{n}{i} \leq 2^n.$$

In the case of equality, we say that a code is *perfect*. A perfect code is a special mathematical object. It is equivalent to a packing of w disjoint spheres of radius e in an n-dimensional binary vector space.

In a perfect code, the quantities w and $\sum_{i=0}^{e} \binom{n}{i}$ must both be powers of 2 (because their product is a power of 2). This turns out to be a severe limitation on the possibilities for existence of perfect codes. Let's ignore the cases $e = 0$ (which means that every binary vector is a codeword), $e = n$ (which means that there is only one codeword), and $e = (n-1)/2$, for n odd (which means that there are only two codewords). Aside from these trivial cases, an infinite family of perfect codes exists with $e = 1$; these are called *Hamming codes*.[3] The only other feasible values of (n, e) are $(23, 3)$ and $(90, 2)$. This is difficult to prove, but a computer search may convince you of its likelihood. Notice that

$$\binom{23}{0} + \binom{23}{1} + \binom{23}{2} + \binom{23}{3} = 2^{11} \quad \text{and} \quad \binom{90}{0} + \binom{90}{1} + \binom{90}{2} = 2^{12}.$$

In fact, there is no perfect code with $(n, e) = (90, 2)$. Assume that there is such a code. Without loss of generality, we may assume that the code contains the all-zero vector of length 90. Let X be the set of binary strings whose first two coordinates are 1 and have exactly one other coordinate equal to 1. Since there are 88 choices for the third coordinate that equals 1, we see that X has 88 elements. Since the code is perfect, each element of X is contained in exactly one sphere of radius 2 around a codeword. Such a codeword must have exactly five 1s (why?). Let Y be the number of codewords that have exactly five 1s. Since each such codeword contains exactly three elements of X, we have $3|Y| = |X| = 88$. But 88 isn't divisible by 3, so we have a contradiction. Therefore, there is no perfect code with $(n, e) = (90, 2)$.

A perfect binary code with $(n, e) = (23, 3)$, called the Golay code, was discovered by the mathematician and physicist Marcel J. E. Golay (1902–1989). We will give a construction of the Golay code based on an idea due to Robert T. Curtis and Tony R. Morris (see [6]). We will produce the Golay code as a vector space, namely, the vector space of linear combinations of the row vectors of a 12×24 matrix. This will yield a code where the codewords have length 24. Deleting any one coordinate results in a perfect code where the codewords have length 23.

Let $\mathcal{G}$ be the icosahedral graph (Figure 5.4). It has twelve vertices, corresponding to the twelve vertices of a regular icosahedron. Two vertices are adjacent (joined by an edge) if and only if the corresponding vertices of the icosahedron are the endpoints of an edge. Assume that the vertices of $\mathcal{G}$ are numbered from 1 to 12. Take vectors of length 24, denoted by $\langle x, y \rangle$, where x is the indicator vector of any subset of vertices of $\mathcal{G}$, and y is the indicator vector of the set of vertices of $\mathcal{G}$ that are nonadjacent to an odd number of vertices

---

[3]Hamming codes were discovered by Richard Hamming (1915–1998).

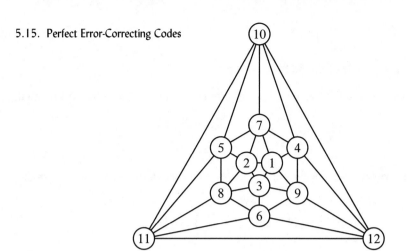

Figure 5.4. The icosahedral graph.

of x. For example, if

$$x = \langle 1, 1, 0, 0, 0, 0, 0, 0, 0, 0, 0, 0 \rangle,$$

then

$$y = \langle 1, 1, 0, 1, 1, 0, 0, 1, 1, 0, 0, 0 \rangle.$$

The generator matrix of the extended Golay code is $M = [I_{12}|A]$, where A is the non-adjacency matrix of the icosahedral graph, and I_{12} is the 12×12 identity matrix. The ij entry of A is 1 if vertices i and j are nonadjacent and 0 if they are adjacent. Every vertex is nonadjacent to itself, so the diagonal entries of A are 1. This matrix M encodes the definition that y represents the set of vertices of $\mathcal{G}$ that are nonadjacent to an odd number of vertices represented by x.

An arbitrary binary vector v of length 12 maps to vM, producing a binary vector of length 24. We obtain 2^{12} different vectors altogether, since the rows of M are linearly independent. If $v = 0$, then vM is the all-zero vector. We claim that the other image vectors have eight 1s, twelve 1s, sixteen 1s, or twenty-four 1s. The easiest way to prove this is by getting a computer to do all the multiplications and keep track of the number of 1s in the image vectors. Deleting one coordinate results in the Golay code with $(n, e) = (23, 3)$, which has 2^{12} codewords of length 23, each pair differing in at least seven coordinates.

The Golay code is special. Its automorphism group (the group of permutations of the coordinates of the codewords that leave the code unchanged) is the Mathieu[4] group M_{23}, a sporadic simple group of order $23 \cdot 22 \cdot 21 \cdot 20 \cdot 16 \cdot 3$. The 759 codewords of weight 8 in the extended Golay code give rise to a Steiner[5] system $S(5, 8, 24)$.

See [45] for a clear account of coding theory.

---

[4]Émile Léonard Mathieu (1835–1890) discovered five sporadic simple groups.
[5]Jakob Steiner (1796–1863) was a geometer who worked primarily in the area of synthetic geometry.

5.16   Binomial Coefficient Magic

In 1891 A. C. Dixon[6] found a formula for the alternating sum of the cubes of binomial coefficients:

$$\sum_{k=0}^{2n}(-1)^k\binom{2n}{k}^3 = (-1)^n\frac{(3n)!}{(n!)^3}.$$

Though there are formulas for the sum of the first and second powers of the binomial coefficients,

$$\sum_{k=0}^{n}\binom{n}{k} = 2^n \quad \text{and} \quad \sum_{k=0}^{n}\binom{n}{k}^2 = \binom{2n}{n}$$

(e.g., see [17]), no formula is known for the sum of the cubes of the binomial coefficients:

$$\sum_{k=0}^{n}\binom{n}{k}^3.$$

Let's prove Dixon's formula. We will look at it as a special case of a more general formula. This is a typical srategy in mathematics. Sometimes a more general formula is easier to prove than a specific case. Since we are dealing with cubes of binomial coefficients, it is fairly natural to sum over products of three different binomial coefficients. An educated guess is the formula called *Dixon's identity*:

$$\sum_{j=-\infty}^{\infty}(-1)^j\binom{a+b}{a+j}\binom{b+c}{b+j}\binom{c+a}{c+j} = \frac{(a+b+c)!}{a!b!c!},$$

where a, b, and c are nonnegative integers. The domain of summation is actually finite, since the summand is 0 outside the interval $-a \le j \le a$. Setting $a = b = c = n$ and making the change of variables $j \leftarrow k - n$, we recover the desired special case.

Let's think of Dixon's identity as an identity in one variable with the other two variables fixed. Thus, we replace a by m, and think of b and c as constants. The identity is

$$f(m) = \sum_{j=-\infty}^{\infty}(-1)^j\binom{m+b}{m+j}\binom{b+c}{b+j}\binom{c+m}{c+j} = \frac{(m+b+c)!}{m!b!c!}.$$

We want to prove this identity for all $m \ge 0$. For $m = 0$, there is only one nonzero summand (corresponding to $j = 0$), and the identity reads $\binom{b+c}{b} = (b+c)!/(b!c!)$, which is true. What happens when m increases to $m + 1$? We have

$$\frac{f(m+1)}{f(m)} = \frac{(m+1+b+c)!}{(m+1)!b!c!} \cdot \frac{m!b!c!}{(m+b+c)!} = \frac{m+b+c+1}{m+1}, \quad m \ge 0.$$

If we can prove this relation, then it will follow by induction that $f(m) = (m+b+c)!/(m!b!c!)$ for all $m \ge 0$, and since we could do the same process for each of the three variables, this will establish Dixon's identity. We write the relation to be proved as

$$(m+1)f(m+1) - (m+1+b+c)f(m) = 0.$$

---

[6]Alfred Cardew Dixon (1865–1936) worked primarily in differential equations. Dixon attributed the summation formula to Frank Morley, the geometer who proved Morley's theorem on the trisectors of a triangle.

We will use telescoping series. Denote the summand in Dixon's identity as

$$f(m, j) = (-1)^j \binom{m+b}{m+j} \binom{b+c}{b+j} \binom{c+m}{c+j}.$$

By definition,

$$\sum_j f(m, j) = f(m).$$

We want to find a function $g(m, j)$ such that

$$(m+1)f(m+1, j) - (m+1+b+c)f(m, j) = g(m, j+1) - g(m, j) \quad (*).$$

If we sum both sides of $(*)$ over all integers j, then the right side will be a telescoping series, and if the telescoping series sums to 0, we are done! Thus, we get a one-step proof if we can find $g(m, j)$.

Experience has shown that we can often choose a function of the form

$$g(m, j) = r(m, j)f(m, j),$$

where $r(m, j)$ is a rational function of m and j. In our case, we can take

$$r(m, j) = \frac{(b+j)(c+j)}{2(j-m-1)}.$$

How do we find such a rational function? One way is to feed some values of $f(m, j)$ into a computer to guess the numerator and denominator, assuming that the degrees are not large. The most important point is that once we have $g(m, j)$, we can easily check (by computer or by hand) that $(*)$ is satisfied. The resulting telescoping series sums to 0, since $f(m, j) = 0$ for j sufficiently large or sufficiently small. Let's prove condition $(*)$ by hand.

In order to cancel lots of factorials, we divide each term of $(*)$ by $f(m, j)$, remembering that $g(m, j) = r(m, j)f(m, j)$. The left side becomes

$$\frac{(m+1)(m+b+1)(m+c+1)}{(m+1+j)(m+1-j)} - (m+b+c+1),$$

and the right side becomes

$$\frac{(b-j)(c-j)}{2(j+m+1)} - \frac{(b+j)(c+j)}{2(j-m-1)}.$$

Algebra shows that the expressions are equal.

This concludes the proof of Dixon's identity and the special sum in our discussion.

The functions $f(m, j)$ and $g(m, j)$ in our solution are called a *WZ pair*, named after the discoverers of the method, Herbert Wilf and Doron Zeilberger. A complete description of the WZ method is given in [42].

We could have used the WZ method directly on Dixon's formula, without the need for the more general identity. However, the rational function would have been more complicated. You might want to use the method to prove the formulas

$$\sum_{k=0}^n \binom{n}{k} = 2^n, \quad \sum_{k=0}^n \binom{n}{k}^2 = \binom{2n}{n}, \quad \text{and} \quad \sum_{k=0}^{2n} (-1)^k \binom{2n}{k}^2 = (-1)^n \binom{2n}{n}.$$

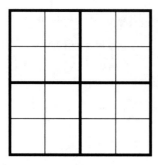

Figure 5.5. A 4 × 4 array.

You can use the method on these formulas directly, without recourse to more general identities.

5.17   A Group of Operations

Let a 4 × 4 array be given (Figure 5.5). Suppose that we can perform three operations on the array:

- Exchange any two rows.
- Exchange any two columns.
- Exchange any two quadrants (the quadrants are outlined with heavy lines).

Combinations of the operations comprise a group. What is its order?

The group is isomorphic to the affine linear group $AG(4, 2)$. This is the group of all transformations

$$v \mapsto vM + t,$$

where M is an invertible 4 × 4 binary matrix, and v and t are binary row vectors of length 4. Its order is

$$(2^4 - 1)(2^4 - 2)(2^4 - 2^2)(2^4 - 2^3)2^4 = 322560.$$

The reason is that there are $2^4 - 1$ choices for the first row of an invertible binary matrix (any binary vector of length 4 except the all 0 vector), $2^4 - 2$ choices for the second row

11	0011	0111	1011	1111
10	0010	0110	1010	1110
01	0001	0101	1001	1101
00	0000	0100	1000	1100
	00	01	10	11

Figure 5.6. A 4 × 4 array with coordinates.

(any binary vector except a multiple of the first row), $2^4 - 2^2$ choices for the third row (any binary vector except a linear combination of the first two rows), $2^4 - 2^3$ choices for the third row (any binary vector except a linear combination of the first three rows), and 2^4 choices for the translation vector t. The order of the group is the product of the numbers of choices.

Let's prove that the group of operations is $AG(4, 2)$.

We put binary coordinates on the cells of the array (Figure 5.6), with the first two coordinates placed in the horizontal direction and the last two placed in the vertical direction. We'll refer to the quadrants of the array as NE, NW, SE, and SW.

Let's look at the action of transvection matrices on the coordinates. A *transvection matrix* is a matrix formed by changing one of the off-diagonal entries of the identity matrix from 0 to 1. For each such matrix M, we consider the result of multiplying on the left by a row vector: vM. In this case, the translation vector t is the all zero vector.

matrix	action
$M_{12} = \begin{bmatrix} 1 & 1 & 0 & 0 \\ 0 & 1 & 0 & 0 \\ 0 & 0 & 1 & 0 \\ 0 & 0 & 0 & 1 \end{bmatrix}$	swaps columns 10 and 11
$M_{21} = \begin{bmatrix} 1 & 0 & 0 & 0 \\ 1 & 1 & 0 & 0 \\ 0 & 0 & 1 & 0 \\ 0 & 0 & 0 & 1 \end{bmatrix}$	swaps columns 01 and 11
$M_{13} = \begin{bmatrix} 1 & 0 & 1 & 0 \\ 0 & 1 & 0 & 0 \\ 0 & 0 & 1 & 0 \\ 0 & 0 & 0 & 1 \end{bmatrix}$	swaps NE and SE quadrants
$M_{31} = \begin{bmatrix} 1 & 0 & 0 & 0 \\ 0 & 1 & 0 & 0 \\ 1 & 0 & 1 & 0 \\ 0 & 0 & 0 & 1 \end{bmatrix}$	swaps NE and NW quadrants
$M_{34} = \begin{bmatrix} 1 & 0 & 0 & 0 \\ 0 & 1 & 0 & 0 \\ 0 & 0 & 1 & 1 \\ 0 & 0 & 0 & 1 \end{bmatrix}$	swaps rows 10 and 11
$M_{43} = \begin{bmatrix} 1 & 0 & 0 & 0 \\ 0 & 1 & 0 & 0 \\ 0 & 0 & 1 & 0 \\ 0 & 0 & 1 & 1 \end{bmatrix}$	swaps rows 01 and 11

In all cases the origin cell, 0000, is fixed.

We will show that these six transvection matrices generate all invertible 4×4 binary matrices. The matrices are labeled M_{ij} so that the single off-diagonal 1 occurs in the ith row and jth column. Every invertible matrix can be written as a product of elementary row operation matrices. The elementary row operations are (1) multiplication by a nonzero scalar, (2) interchange of two rows, and (3) replacement of a row by that row plus a scalar multiple of another row. With operations over a two-element field, scalar multiplication doesn't amount to much, since the only nonzero scalar is 1. Transvection matrices correspond to elementary row operations of type (3). We assume that the operating matrix is on the left and the operated-upon matrix is on the right. For example, multiplying a 4×4 matrix A on the left by M_{12} replaces the first row of A by the sum of the first and second rows. In general, the product $M_{ij} A$ is the matrix in which row i of A is replaced by the sum of rows i and j. The type (2) elementary row operations are easy to form with our transvection matrices. For instance, the matrix product $M_{12} M_{21} M_{12} A$ is the matrix in which the first two rows of A are interchanged. With permutation matrices at our disposal, it is easy to generate the other transvection matrices. How would you generate the matrix M_{23}?

Thus, the six matrices generate all 4×4 invertible binary matrices. Addition of a nonzero translation vector t moves the origin cell. Therefore, the group of operations is the same as $AG(4, 2)$.

6

Elegant Solutions

The essence of mathematics resides in its freedom.

—GEORG CANTOR (1845–1918)

What makes a difficult mathematics problem easy? Sometimes there is a sudden flash of understanding. Sometimes past experience points out the right direction to take. This chapter presents problems whose solutions illustrate concepts or techniques that can be appreciated for their power and beauty, and may be useful to you in future problem solving.

6.1 A Tetrahedron and Four Spheres

Four spheres of radius 1 are contained in a regular tetrahedron in such a way that each is tangent to three faces of the tetrahedron and to the other three spheres. What is the side length of the tetrahedron?

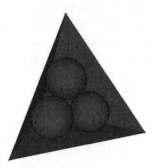

In a regular tetrahedron, let r be the ratio of the distance between its center and a face to the side length. Line segments joining the four centers of the mutually tangent spheres form a small regular tetrahedron of side length 2. The small tetrahedron and the circumscribing tetrahedron have the same center, and their faces are one unit apart (the radii of the spheres). Hence, the side length of the circumscribing tetrahedron is $(2r + 1)/r$. We will find r and thereby find this length.

Given a regular tetrahedron of side length 1, the altitude of a face has length $\sqrt{3}/2$ (by the Pythagorean theorem). Since the center of an equilateral triangle divides the altitudes in the ratio 2 : 1, the distance between the center of a face and a vertex is $\sqrt{3}/3$ and hence (again by the Pythagorean theorem) the altitude of the tetrahedron is $\sqrt{6}/3$. Since the center of a regular tetrahedron divides the altitudes in the ratio 3 : 1, we conclude that $r = \sqrt{6}/12$ and therefore the side length of the circumscribing tetrahedron is $2 + 2\sqrt{6}$.

6.2 Alphabet Cubes

Suppose that you have 27 wooden cubes, all the same size. On each face of the first cube is the letter A. On each face of the second cube is the letter B, and so on, all the way to Z. The twenty-seventh cube is blank. Is it possible to arrange the cubes in a $3 \times 3 \times 3$ cube, with the blank cube in the middle, so that consecutive letters of the alphabet occur on cubes sharing a face?

It is impossible to accomplish this. Consider a checkerboard coloring of the cubes, using red and black, so that cubes that share a face have different colors. If there were such an arrangement, then as we go through the alphabet, we would change cube colors at each step. But this would mean that we have thirteen cubes of each color. However, the checkerboard coloring has fourteen cubes of one color and thirteen of the other. Omitting the middle cube makes the distribution of colors even more imbalanced, with fourteen of one color and twelve of the other color. Therefore, our desired path through the alphabet is impossible.

It is possible to solve the problem if consecutive alphabetic cubes have an edge or a face in common, and we can even make it so that this applies to A and Z.

H	G	E
I	F	D
A	B	C

T	U	V
J		W
Z	Y	X

S	R	Q
K	M	P
L	N	O

6.3 A Triangle in an Ellipse

Show how to inscribe a triangle of maximum area in an ellipse.

What happens when the ellipse is a circle? You may conjecture that a triangle of maximum area inscribed in a circle is equilateral. Let's prove this by simple geometry. Suppose that $\triangle ABC$ is a triangle of maximum area inscribed in a circle. We will show that it is equilateral. Let a line parallel to AB be tangent to the circle at a point C'. There are two

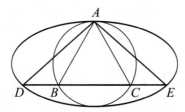

Figure 6.1. Inscribing a maximum-area triangle in an ellipse.

choices for C'; pick the one so that the distance between AB and C' is the greatest. If C is any point on the circle other than C', the area of $\triangle ABC$ will be less than the area of $\triangle ABC'$, since the altitude of the first triangle will be less than the altitude of the second triangle. Because $\triangle ABC$ is a maximum-area inscribed triangle, $C = C'$. Since C lies on the perpendicular bisector of AB, which passes through the center of the circle, we conclude that $AC = BC$. By a similar argument, $BC = BA$, and therefore $\triangle ABC$ is an equilateral triangle.

Now that we know that a maximum-area triangle inscribed in a circle is an equilateral triangle, we can show how to find a maximum-area triangle inscribed in an ellipse. Without loss of generality, suppose that the ellipse has the equation

$$\frac{x^2}{a^2} + \frac{y^2}{b^2} = 1,$$

or

$$\frac{b^2 x^2}{a^2} + y^2 = b^2.$$

By the change of coordinates $x' = bx/a$, $y' = y$, the equation becomes

$$x'^2 + y'^2 = b^2,$$

a circle of radius b. Thus, an ellipse may be thought of as a circle stretched by a constant factor. This scales the area of a plane figure by the factor. Hence, a maximum-area triangle inscribed in the circle is stretched to produce a maximum-area triangle inscribed in the ellipse.

Figure 6.1 shows the construction. In the diagram, the ellipse lies outside the circle, but it could just as well be the other way around.

We place A at an intersection point of the circle and ellipse, and let B and C be the points on the circle forming the equilateral triangle $\triangle ABC$. We draw the line containing B and C, which intersects the ellipse at D and E. Then $\triangle ADE$ is a maximum-area triangle inscribed in the ellipse. Maximum-area triangles inscribed in ellipses come in lots of shapes. We can stretch any equilateral triangle inscribed in the circle.

6.4   About the Roots of a Cubic

The cubic polynomial

$$x^3 + 11x^2 + 19x - 100$$

has three roots, one real and two complex. What is the sum of their squares?

We could find the roots explicitly, square them, and add, but we will be more elegant and relate the roots to the polynomial's coefficients. A monic cubic polynomial with roots r_1, r_2, and r_3 can be factored as

$$(x - r_1)(x - r_2)(x - r_3) = x^3 - (r_1 + r_2 + r_3)x^2 + (r_1 r_2 + r_1 r_3 + r_2 r_3)x - r_1 r_2 r_3.$$

Equating coefficients, we see that[1]

$$r_1 + r_2 + r_3 = -11,$$

$$r_1 r_2 + r_1 r_3 + r_2 r_3 = 19,$$

$$r_1 r_2 r_3 = 100.$$

Therefore, the sum of the squares of the roots is

$$r_1^2 + r_2^2 + r_3^2 = (r_1 + r_2 + r_3)^2 - 2(r_1 r_2 + r_1 r_3 + r_2 r_3)$$

$$= (-11)^2 - 2 \cdot 19$$

$$= 83.$$

Here is a method to calculate the sum of the kth powers of the roots of our polynomial, for any positive integer k. Let

$$p_k = r_1^k + r_2^k + r_3^k, \quad k \geq 0.$$

Each root satisfies the equation

$$x^3 = -11x^2 - 19x + 100,$$

the characteristic equation of the recurrence relation

$$p_k = -11 p_{k-1} - 19 p_{k-2} + 100 p_{k-3}, \quad k \geq 3.$$

Hence, the sequence $\{p_k\}$ is given by this recurrence relation and the initial conditions

$$p_0 = 3, \quad p_1 = -11, \quad p_2 = 83.$$

It's now a simple matter to use the recurrence formula to calculate the sum of the kth powers of the roots. For example, the sum of the cubes of the roots is

$$p_3 = -11(83) - 19(-11) + 100(3) = -404.$$

What is the sum of the fourth powers of the roots? The answer is the year that Wolfgang Amadeus Mozart (1756–1791) wrote the opera *Apollo et Hyacinthus*.

---

[1]These relations are called Viète's formulas, named after François Viète [Vieta] (1540–1603).

6.5   Distance on Planet X

Planet X is a sphere of radius 1000 km. On Planet X, a city is located at 10° degrees N latitude and 20° degrees E longitude, while another city is located at 30° N latitude and 60° degrees E longitude. What is the shortest distance between the two cities along the surface of Planet X?

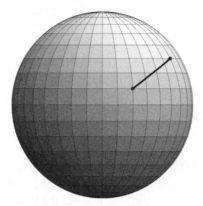

   Nothing is special about the coordinates given for the two cities, so we want a method for finding the minimum distance along the surface of Planet X between any two points. A shortest path between two points on a sphere is called a *geodesic*. A geodesic is an arc of a great circle, so if we know the central angle of the great circle containing the points, we can multiply by the radius to get its length. Latitude and longitude are essentially spherical coordinates. If we convert them to Cartesian coordinates of vectors v and w, then we can use the dot product formula

$$\cos \Omega = \frac{v \cdot w}{|v||w|}$$

to find the central angle Ω separating the two cities, and multiply Ω (in radians) by the radius to find the length of the geodesic.

   We take the spherical coordinates for a point on the surface of a planet with radius r to be (r, ϕ, θ), where ϕ is the latitude and θ is the longitude of the point. The corresponding Cartesian coordinates are

$$(x, y, z) = (r \cos \phi \cos \theta, r \cos \phi \sin \theta, r \sin \phi).$$

Let the two points have spherical coordinates (r, ϕ_1, θ_1) and (r, ϕ_2, θ_2) and hence Cartesian coordinates

$$v = (r \cos \phi_1 \cos \theta_1, r \cos \phi_1 \sin \theta_1, r \sin \phi_1) \quad \text{and}$$

$$w = (r \cos \phi_2 \cos \theta_2, r \cos \phi_2 \sin \theta_2, r \sin \phi_2).$$

By the dot product formula, the cosine of the central angle ($\cos \Omega$) is

$$\cos \phi_1 \cos \phi_2 \cos \theta_1 \cos \theta_2 + \cos \phi_1 \cos \phi_2 \sin \theta_1 \sin \theta_2 + \sin \phi_1 \sin \phi_2,$$

and the length of the geodesic is

$$r \cos^{-1}(\cos \phi_1 \cos \phi_2 \cos \theta_1 \cos \theta_2 + \cos \phi_1 \cos \phi_2 \sin \theta_1 \sin \theta_2 + \sin \phi_1 \sin \phi_2).$$

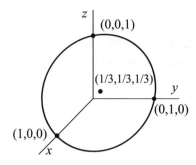

Figure 6.2. A tilted circle.

In our problem, $r = 1000$ km, $\phi_1 = 10°$, $\theta_1 = 20°$, $\phi_2 = 30°$, and $\theta_2 = 60°$. It follows that the two alien cities are separated by a central angle of approximately $\Omega \doteq 0.74$ radians ($42°$) and a distance of approximately 740 km.

We can use the dot product method to prove a formula of spherical trigonometry known as the *spherical law of cosines*. Consider a spherical triangle on a unit sphere, with angles A, B, C and opposite sides a, b, and c. The sides of the triangle are arcs of great circles of the sphere. The angles are determined by the planes of the great circles defining the sides. Since the sphere has radius 1, the arcs subtend central angles a, b, and c. Set up a Cartesian coordinate system so that the xy-plane passes through A, C, and the center of the sphere. Choose the x-axis so that C has coordinates $(1, 0, 0)$. By plane trigonometry, A has coordinates $(\cos b, \sin b, 0)$ and B has coordinates $(\cos a, \sin a \cos C, \sin a \sin C)$. The dot product formula gives us the spherical law of cosines:

$$\cos c = \cos a \cos b + \sin a \sin b \cos C.$$

6.6   A Tilted Circle

Find parametric equations for the circle in $\mathbf{R}^3$ that passes through the points $(1, 0, 0)$, $(0, 1, 0)$, and $(0, 0, 1)$. See Figure 6.2. Parametric equations are equations for x, y, z in terms of a new variable t. They allow us to trace the circle in $\mathbf{R}^3$ as a vector function of one variable.

Sometimes a mathematical problem can be solved by taking a hint from physics. The unit circle centered at the origin in $\mathbf{R}^2$ is given by the parametric equations

$$x = \cos t,$$

$$y = \sin t, \qquad 0 \le t < 2\pi.$$

The components of the velocity of a point moving around the circle at constant speed are

$$x' = -\sin t,$$

$$y' = \cos t, \qquad 0 \le t < 2\pi.$$

The variable t, for time, is defined so that the point moves around the circle once every 2π

units of time. Acceleration of the point is given by the second derivatives:

$$x'' = -\cos t,$$

$$y'' = -\sin t, \qquad 0 \leq t < 2\pi.$$

The acceleration vector is the opposite of the position vector:

$$x'' = -x, \quad y'' = -y.$$

This applies to any circular motion at constant normalized speed, where the circle is centered at the origin. In three dimensions, the position function $(x(t), y(t), z(t))$ satisfies

$$x'' = -x, \quad y'' = -y, \quad z'' = -z.$$

Solutions to these equations constitute *simple harmonic motion*, a linear combination of sine and cosine functions. Thus

$$x = A_1 \cos t + B_1 \sin t, \quad y = A_2 \cos t + B_2 \sin t, \quad z = A_3 \cos t + B_3 \sin t,$$

for some constants $A_1, B_1, A_2, B_2, A_3, B_3$.

In our problem, the center of the circle is $(1/3, 1/3, 1/3)$ as this point is equidistant from $(1, 0, 0)$, $(0, 1, 0)$, and $(0, 0, 1)$, and also coplanar with them (since the sum of its coordinates is 1). The radius of the circle is the distance between its center and $(1, 0, 0)$, that is, $\sqrt{6}/3$, but we don't need this value.

We add the center to our parametric equations thus far:

$$x = \frac{1}{3} + A_1 \cos t + B_1 \sin t$$

$$y = \frac{1}{3} + A_2 \cos t + B_2 \sin t$$

$$z = \frac{1}{3} + A_3 \cos t + B_3 \sin t, \qquad 0 \leq t < 2\pi.$$

Given that $x(0) = 1$, we find $A_1 = 2/3$, and from $x(2\pi/3) = 0$, we find $B_1 = 0$. So

$$x = \frac{1}{3} + \frac{2}{3} \cos t.$$

By symmetry,

$$x = \frac{1}{3} + \frac{2}{3} \cos t$$

$$y = \frac{1}{3} + \frac{2}{3} \cos \left(t - \frac{2\pi}{3} \right)$$

$$z = \frac{1}{3} + \frac{2}{3} \cos \left(t - \frac{4\pi}{3} \right), \qquad 0 \leq t < 2\pi.$$

The circle passes through the points $(1, 0, 0)$, $(0, 1, 0)$, and $(0, 0, 1)$ when $t = 0, 2\pi/3$, and $4\pi/3$, respectively.

Another problem that can be solved by appealing to physical intuition is the construction of the Fermat point of a triangle [26, pp. 34–36]. The book [30] is devoted to the method of using physical reasoning to solve math problems.

6.7 The Millionth Fibonacci Number

What are the first three digits of $F_{1000000}$, the one millionth Fibonacci number? The Fibonacci sequence $\{F_n\}$ is defined by

$$F_0 = 0, \ F_1 = 1,$$

$$F_n = F_{n-1} + F_{n-2}, \quad n \geq 2.$$

It isn't practical to compute $F_{1000000}$, even with a computer, because it has too many digits. But we are only asked to find the first three digits.

We use an explicit formula for the nth Fibonacci number:

$$F_n = \frac{\phi^n - \hat{\phi}^n}{\sqrt{5}}, \quad n \geq 0,$$

where

$$\phi = \frac{1 + \sqrt{5}}{2}, \quad \hat{\phi} = \frac{1 - \sqrt{5}}{2}.$$

The constant ϕ is the golden ratio described in Chapter 1. Since $\phi \doteq 1.6$ and $\hat{\phi} \doteq -0.6$, the Fibonacci sequence grows like the exponential sequence $\{\phi^n / \sqrt{5}\}$. The difference between the two sequences becomes exponentially small as n tends to infinity, and is therefore negligible. Thus

$$F_n \doteq \frac{\phi^n}{\sqrt{5}}.$$

Taking the base-10 logarithm,

$$\log F_n \doteq n \log \phi - \log \sqrt{5}.$$

For $n = 1000000$, we have

$$\log F_{1000000} \doteq 1000000 \log \phi - \log \sqrt{5} \doteq 208987.2908.$$

We see that the number of digits in the millionth Fibonacci number is 208,988. To get the first three digits, we compute

$$10^{0.2908} \doteq 1.953,$$

so the first three digits of the millionth Fibonacci number are 195. In scientific notation, the millionth Fibonacci number is

$$F_{1000000} \doteq 1.95 \times 10^{208987}.$$

It isn't difficult to find the last three digits of the millionth Fibonacci number. We can run the Fibonacci recurrence relation, keeping only the last three digits at each step. The millionth Fibonacci number ends in 875. The last three digits of Fibonacci numbers repeats every 1500 terms. Since F_{1000} ends in 875, and 1500 divides $1000000 - 1000$, this confirms that the last three digits of $F_{1000000}$ are 875.

6.8   The End of a Conjecture

Let σ be a permutation of the set $\{1, \ldots, n\}$, for some positive integer n. The *order* of σ is the smallest positive integer k such that σ^k (σ applied k times) is the identity permutation. A natural question is, what is the greatest possible order of σ?

The order of a permutation is the least common multiple of the lengths of its disjoint cycles. Experimentation shows that the greatest possible order of a permutation of 10 elements is $2 \cdot 3 \cdot 5 = 30$, which occurs for permutations consisting of disjoint cycles of lengths 2, 3, and 5. Based on this and other small examples, we could make a conjecture:

Conjecture.   If n is the sum of consecutive prime numbers, $n = 2 + 3 + \cdots + p$, then the greatest possible order of a permutation of the set $\{1, \ldots, n\}$ is $2 \cdot 3 \cdot \cdots \cdot p$.

This conjecture is plausible but false. Disprove it.

To disprove a conjecture, it suffices to find an instance when it doesn't hold. How do we do this in the present case? We will start with a partition $n = 2 + 3 + \cdots + p$, for some p, and replace some of the primes by powers of these primes, in such a way that the sum of the terms is still n. Since the prime powers will have no common factors, we can hope that their least common multiple will be greater than the least common multiple of the terms that they replace. In fact, the conjecture is true for all primes p up to 19. So let's look at the case $p = 23$. Consider the partition

$$100 = 2 + 3 + 5 + 7 + 11 + 13 + 17 + 19 + 23.$$

Replace the 2, 3, and 23 by 16, 9, 1, 1, and 1 (retaining the sum). Then the ratio of the least common multiple of the numbers in the second partition to the least common multiple of the numbers in the first partition is

$$\frac{16 \cdot 9}{2 \cdot 3 \cdot 23} = \frac{144}{138},$$

which is greater than 1. Hence, a permutation of 100 whose cycle lengths are the terms of the new partition will have a greater order than one given by the partition in the conjecture.

The function $g(n)$ that gives the greatest possible order of a permutation of n elements is called *Landau's function*, named after Edmund Landau (1877–1938). There is no simple formula known for $g(n)$, although some of its properties are known. For instance, $\ln g(n)$ is asymptotic to $\sqrt{n \ln n}$, which means that

$$\lim_{n \to \infty} \frac{\ln g(n)}{\sqrt{n \ln n}} = 1.$$

A surprising characteristic of $g(n)$ is that it is constant for arbitrarily many consecutive values of n. See [33] for a very readable account of Landau's function.

6.9   A Zero-Sum Game

Consider a game in which two players, A and B, choose one of two alternatives, x and y. Based on their choices, there is a payoff from player B to player A according to the following table.

		B	
		x	y
A	x	$+3$	-2
	y	-1	$+2$

For example, if A and B both choose x, then B gives 3 points to A. Suppose that A and B play this game repeatedly, at each turn randomly choosing x and y according to fixed probabilities p_1, p_2, q_1, q_2, as shown in the table. What is the expected long-term outcome of the game?

Let Ω be the expected payoff to player A. Thus

$$\Omega = 3p_1q_1 - 2p_1q_2 - p_2q_1 + 2p_2q_2.$$

Both A and B are trying to maximize their expected gain. Player A should choose probabilities p_i that maximize Ω no matter what probabilities q_j player B chooses. At the same time, player B should choose probabilities q_j that minimize Ω no matter what probabilities p_i player A chooses. This fundamental principle of zero-sum games is summarized by the *equilibrium formula*

$$\max_{p_i} \min_{q_i} \Omega = \min_{q_i} \max_{p_i} \Omega.$$

Geometrically, this min-max value is the saddle point of the surface given by Ω in three-dimensional space where the independent variables are $p = p_1$ and $q = q_1$. Let's calculate it. We have

$$\Omega(p,q) \;=\; 3pq - 2p(1-q) - (1-p)q + 2(1-p)(1-q)$$
$$\;=\; 8pq - 4p - 3q + 2.$$

At a critical point, the partial derivatives are $\partial\Omega/\partial p = 8q - 4 = 0$, so that $q = 1/2$, and $\partial\Omega/\partial q = 8p - 3 = 0$, so that $p = 3/8$. The determinant of the Hessian matrix is

$$\begin{vmatrix} \dfrac{\partial^2\Omega}{\partial p^2} & \dfrac{\partial^2\Omega}{\partial p\,\partial q} \\[2ex] \dfrac{\partial^2\Omega}{\partial q\,\partial p} & \dfrac{\partial^2\Omega}{\partial q^2} \end{vmatrix} = \begin{vmatrix} 0 & 8 \\ 8 & 0 \end{vmatrix} = -64 < 0.$$

This confirms that the critical point is a saddle point, as indicated in Figure 6.3. The payoff for A is $\Omega(3/8, 1/2) = 1/2$. Therefore, B should stop playing this game.

The min-max equilibrium result for zero-sum games was formulated by John von Neumann (1903–1957).

If the payoffs are λ_{11}, λ_{12}, λ_{21}, and λ_{22}, as shown below, can you determine the condition on the λ's that guarantees the existence of a min-max solution?

		B	
		x	y
A	x	λ_{11}	λ_{12}
	y	λ_{21}	λ_{22}

An excellent reference on game theory, including zero-sum games, is [50].

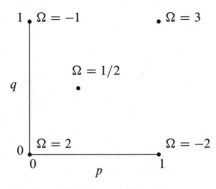

Figure 6.3. The payoffs in a zero-sum game.

6.10 An Expected Maximum

A man has a bottle of vitamin pills from which he takes a half pill per day. Each day, he selects a pill at random from the bottle. If it is a whole pill he cuts it in half, takes a half pill, and puts the other half back in the bottle. If it is a half pill, he takes it. He continues this daily regimen until the bottle is empty. If the bottle starts with n pills, show that the expected maximum number of half pills in the bottle tends to n/e as n tends to infinity. (The savvy problem-solver won't be surprised by the appearance of e in a probability problem.)

We will set up and solve a system of differential equations. Let $p(t)$ be the expected number of whole pills and $h(t)$ the expected number of half pills on day t, where $0 \le t \le 2n$. The probability that a whole pill is selected on day t is $p(t)/(p(t) + h(t))$ and the probability that a half pill is selected is $h(t)/(p(t) + h(t))$. If it is a whole pill, the number of whole pills goes down by 1 while the number of half pills goes up by 1. If it is a half pill, the number of whole pills is unchanged while the number of half pills goes down by 1. These observations give rise to a system of difference equations:

$$p(t + 1) = p(t) - \frac{p(t)}{p(t) + h(t)}$$

$$h(t + 1) = h(t) + \frac{p(t)}{p(t) + h(t)} - \frac{h(t)}{p(t) + h(t)}, \quad 0 \le t \le 2n - 1,$$

where $p(0) = n$ and $h(0) = 0$.

We replace this discrete process with a continuous one, i.e., a system of differential equations:

$$\frac{dp}{dt} = \frac{-p}{p + h}$$

$$\frac{dh}{dt} = \frac{p - h}{p + h},$$

where p and h are functions of t, with $p(0) = n$ and $h(0) = 0$. This approximation to a discrete system by a continuous one becomes better as n increases.

We see that dh/dt is positive until $h = p$. Therefore, the maximum value of h occurs when $h = p$. Dividing the two differential equations gives the single differential equation

$$\frac{dh}{dp} = \frac{h - p}{p},$$

where h is a function of p, and $h = 0$ when $p = n$. By inspection,

$$h = p \ln(n/p),$$

which, with $p = h$, yields

$$h = n/e.$$

Thus, as n increases, the expected maximum number of half pills approaches n/e.

What do the curves $p(t)$ and $h(t)$ look like? Since a whole pill is consumed in two days (not necessarily consecutive) and a half pill is consumed in one day,

$$2p + h + t = 2n,$$

and hence

$$t = 2n - 2p - p \log(n/p).$$

(It's impossible to write p and h in terms of elementary functions of t.) The maximum value of h occurs when $p = n/e$, and this gives

$$t = (2 - 3/e)n.$$

The curves $p(t)$ and $h(t)$ are shown below.

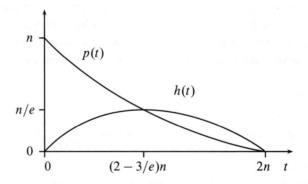

6.11   Walks on a Graph

A *graph* is a set of points and a collection of edges joining pairs of points. A *walk* on a graph is a sequence of vertices such that consecutive vertices are adjacent (joined by an edge). Figure 6.4 depicts a graph with four vertices and three edges. An example of a walk on this graph is the sequence 1, 2, 1, 2, 3. The length of a walk is the number of edges traversed in the graph. The walk 1, 2, 1, 2, 3 has length 4.

Figure 6.4. A graph to walk on.

Let $a_{ij}^{(n)}$ be the number of walks of length n from vertex i to vertex j, where $n \geq 1$ and $1 \leq i, j \leq 4$. Let us a formula for $a_{ij}^{(n)}$.

To solve this problem, we introduce the *adjacency matrix* of the graph:

$$A = \begin{bmatrix} 0 & 1 & 0 & 0 \\ 1 & 0 & 1 & 0 \\ 0 & 1 & 0 & 1 \\ 0 & 0 & 1 & 0 \end{bmatrix}.$$

The ij entry of A is 1 if there is an edge from vertex i to vertex j, and 0 otherwise. By definition, A is a symmetric matrix.

One of the useful properties of an adjacency matrix is that its square tells us the number of walks of length 2 between any two vertices:

$$A^2 = \begin{bmatrix} 1 & 0 & 1 & 0 \\ 0 & 2 & 0 & 1 \\ 1 & 0 & 2 & 0 \\ 0 & 1 & 0 & 1 \end{bmatrix}.$$

For example, the 12 entry of A^2 is 0, and there are no walks of length 2 from vertex 1 to vertex 2. Why does the ij entry of A^2 equal the number of walks of length 2 from i to j? Consider what happens when we form the matrix product A^2. Since

$$A = \begin{bmatrix} a_{11}^{(1)} & a_{12}^{(1)} & a_{13}^{(1)} & a_{14}^{(1)} \\ a_{21}^{(1)} & a_{22}^{(1)} & a_{23}^{(1)} & a_{24}^{(1)} \\ a_{31}^{(1)} & a_{32}^{(1)} & a_{33}^{(1)} & a_{34}^{(1)} \\ a_{41}^{(1)} & a_{42}^{(1)} & a_{43}^{(1)} & a_{44}^{(1)} \end{bmatrix},$$

the ij entry of A^2 is

$$a_{i1}^{(1)}a_{1j}^{(1)} + a_{i2}^{(1)}a_{2j}^{(1)} + a_{i3}^{(1)}a_{3j}^{(1)} + a_{i4}^{(1)}a_{4j}^{(1)}.$$

Each term $a_{ik}^{(1)}a_{kj}^{(1)}$ is the product of two numbers equal to 0 or 1, which is non-zero if and only if $a_{ik}^{(1)} = 1$ and $a_{kj}^{(1)} = 1$, that is, the vertices i and k are adjacent and the vertices k and j are adjacent. This happens precisely when there is a walk of length 2 from i to j through k. Since we sum over all vertices k, the ij entry of A^2 is $a_{ij}^{(2)}$.

For $n \geq 1$, the ij entry of A^n is $a_{ij}^{(n)}$, the number of walks of length n from i to j, which can be proved by mathematical induction.

Let's investigate some higher powers of A. We have

$$A^3 = \begin{bmatrix} 0 & 2 & 0 & 1 \\ 2 & 0 & 3 & 0 \\ 0 & 3 & 0 & 2 \\ 1 & 0 & 2 & 0 \end{bmatrix},$$

$$A^4 = \begin{bmatrix} 2 & 0 & 3 & 0 \\ 0 & 5 & 0 & 3 \\ 3 & 0 & 5 & 0 \\ 0 & 3 & 0 & 2 \end{bmatrix},$$

and

$$A^5 = \begin{bmatrix} 0 & 5 & 0 & 3 \\ 5 & 0 & 8 & 0 \\ 0 & 8 & 0 & 5 \\ 3 & 0 & 5 & 0 \end{bmatrix}.$$

It appears that the Fibonacci numbers are involved. The Fibonacci numbers F_n are defined by the recurrence formula

$$F_0 = 1, \ F_1 = 1, \quad F_n = F_{n-1} + F_{n-2}, \ n \geq 2,$$

so

$$\{F_n\} = \{0, \ 1, \ 1, \ 2, \ 3, \ 5, \ 8, \ 13, \ 21, \ 34, \ \ldots\}.$$

We claim that

$$A^n = \begin{cases} \begin{bmatrix} F_{n-1} & 0 & F_n & 0 \\ 0 & F_{n+1} & 0 & F_n \\ F_n & 0 & F_{n+1} & 0 \\ 0 & F_n & 0 & F_{n-1} \end{bmatrix}, & \text{for } n \text{ even} \\[3em] \begin{bmatrix} 0 & F_n & 0 & F_{n-1} \\ F_n & 0 & F_{n+1} & 0 \\ 0 & F_{n+1} & 0 & F_n \\ F_{n-1} & 0 & F_n & 0 \end{bmatrix}, & \text{for } n \text{ odd.} \end{cases}$$

We have verified this for $n = 1$ and $n = 2$. Suppose that the formula holds for an even value of n. Then

$$A^{n+1} = \begin{bmatrix} F_{n-1} & 0 & F_n & 0 \\ 0 & F_{n+1} & 0 & F_n \\ F_n & 0 & F_{n+1} & 0 \\ 0 & F_n & 0 & F_{n-1} \end{bmatrix} \begin{bmatrix} 0 & 1 & 0 & 0 \\ 1 & 0 & 1 & 0 \\ 0 & 1 & 0 & 1 \\ 0 & 0 & 1 & 0 \end{bmatrix}$$

$$= \begin{bmatrix} 0 & F_{n-1} + F_n & 0 & F_n \\ F_{n+1} & 0 & F_{n+1} + F_n & 0 \\ 0 & F_n + F_{n+1} & 0 & F_{n+1} \\ F_n & 0 & F_n + F_{n-1} & 0 \end{bmatrix}$$

$$= \begin{bmatrix} 0 & F_{n+1} & 0 & F_n \\ F_{n+1} & 0 & F_{n+2} & 0 \\ 0 & F_{n+2} & 0 & F_{n+1} \\ F_n & 0 & F_{n+1} & 0 \end{bmatrix}.$$

This is the correct formula for $n + 1$. Similarly, we can show that if the formula holds for an odd value of n, then it holds for $n + 1$. It follows by mathematical induction that the formula holds for all $n \geq 1$. The formula gives a simple way to find $a_{ij}^{(n)}$.

Figure 6.5. A 2×3 grid.

6.12   Rotations of a Grid

Let a 2×3 grid be given, containing the integers 1 through 6, as in Figure 6.5.

Let L (for "left") be the operation of rotating the left-most 2×2 sub-grid $90°$ clockwise, and R (for "right") be the operation of rotating the right-most 2×2 sub-grid $90°$ clockwise. That is, L is the permutation $(1, 2, 5, 4)(3)(6)$ and R is the permutation $(2, 3, 6, 5)(1)(4)$. Given successive applications of these two operations, how many different permutations of the grid can result? What is the group?

We are faced with two questions: how many elements are in the group and what is its structure? To start, we look at some combinations of group elements.

Using the rotations L and R, we can put any of the numbers in any cell of the grid. Let's say that we put a selected number in the $(1, 1)$ position of the grid. Then we can put any of the remaining numbers in the $(2, 1)$ position, using powers of $L^2 R L^{-1}$, which fixes the $(1, 1)$ position and moves the other cells in a 5-cycle, and any of the remaining entries in the $(1, 2)$ position using powers of R. Thus, we see that there are at least $6 \cdot 5 \cdot 4 = 120$ permutations of the grid.

Because $120 = 5!$, we could guess that the group is isomorphic to the group of permutations of a five-element set, for this group has order $5!$. This group, denoted by S_5, is called the *symmetric group* of degree 5. We will prove this is correct.

To prove that the group is isomorphic to S_5, we are confronted with the question, what five things can we permute? In an effort to find a five somewhere in the problem, we notice that the number of pairs of cells of the grid is $\binom{6}{2} = 15$, so perhaps we should put the pairs of cells into five groups of three each, as

$$\mathcal{A} = \{\{1, 4\}, \{2, 6\}, \{3, 5\}\}$$

$$\mathcal{B} = \{\{1, 3\}, \{2, 5\}, \{4, 6\}\}$$

$$\mathcal{C} = \{\{1, 5\}, \{2, 4\}, \{3, 6\}\}$$

$$\mathcal{D} = \{\{1, 2\}, \{3, 4\}, \{5, 6\}\}$$

$$\mathcal{E} = \{\{1, 6\}, \{2, 3\}, \{4, 5\}\}.$$

It's helpful to picture the sets with lines representing the pairs of cells. Here is the picture for the set $\mathcal{A}$.

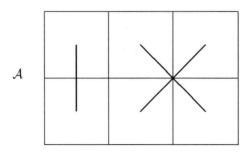

The operations L and R move each set $\mathcal{A}$, $\mathcal{B}$, $\mathcal{C}$, $\mathcal{D}$, and $\mathcal{E}$ to another such set. That is, the sets are permuted by L and R. Specifically,

$$L = (\mathcal{A}, \mathcal{D}, \mathcal{B}, \mathcal{E})(\mathcal{C}) \quad \text{and} \quad R = (\mathcal{A})(\mathcal{B}, \mathcal{E}, \mathcal{C}, \mathcal{D}).$$

At this point we know that the group is a subgroup of S_5. We calculate

$$LR = (\mathcal{A}, \mathcal{B}, \mathcal{C}, \mathcal{D}, \mathcal{E}),$$

and

$$RLR = (\mathcal{A}, \mathcal{B}, \mathcal{C})(\mathcal{D}, \mathcal{E}).$$

Applying RLR three times, the 3-cycle disappears and we are left with a transposition:

$$(RLR)^3 = (\mathcal{A})(\mathcal{B})(\mathcal{C})(\mathcal{D}, \mathcal{E}).$$

Hence, the group contains a 5-cycle and a transposition of adjacent terms in the 5-cycle ($\mathcal{D}$ and $\mathcal{E}$). It is well known (e.g., see [25, p. 118]) that they generate S_5.

Let's show that the transposition $(1, 2)$ and the cycle $(1, 2, \ldots, n)$ generate all permutations of $\{1, 2, \ldots, n\}$. We'll demonstrate this in the case $n = 5$ but the same argument works in general. We can generate every transposition of consecutive numbers. The technique is *conjugation*, which sends an element x to a new element $g^{-1}xg$. Thus

$$(5, 4, 3, 2, 1)(1, 2)(1, 2, 3, 4, 5) = (2, 3)(1)(4)(5),$$

and we have generated the transposition $(2, 3)$. Continuing, we generate $(3, 4)$, $(4, 5)$, and $(5, 1)$. Now that we have transpositions of consecutive numbers, we can generate all transpositions. For instance,

$$(2, 3)(3, 4)(4, 5)(3, 4)(2, 3) = (2, 5).$$

So we have the transposition $(2, 5)$. As an exercise, show how to obtain the transposition $(3, 5)$. Now that we have all transpositions, it is easy to obtain any cycle. For instance,

$$(3, 5)(3, 1)(3, 4) = (3, 5, 1, 4).$$

Finally, since all permutations are products of cycles, we are done.

An alternative presentation of the group is worth considering. We relabel the entries of the grid.

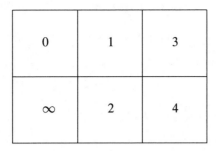

The entries are the residue classes of integers modulo 5, i.e., the numbers 0, 1, 2, 3, and 4, together with ∞. We define two functions

$$f(x) = \frac{1}{2x+1}, \quad g(x) = 3x.$$

You can verify that f represents the operation L (it rotates the left-most 2×2 sub-grid by $90°$), and g represents R (it rotates the right-most 2×2 sub-grid by $90°$). Remember to reduce each value of the functions modulo 5. For example,

$$f: 0 \mapsto 1 \mapsto \frac{1}{3} = \frac{2}{6} = \frac{2}{1} = 2 \mapsto \frac{1}{5} = \frac{1}{0} = \infty \mapsto \frac{1}{\infty} = 0.$$

The resulting group of compositions of f and g is called the group of *linear fractional transformations* of the five-element field together with ∞.

The problem can be generalized to allow any rotation of any 2×2 sub-grid of an $m \times n$ grid, where $2 \le m \le n$. The group is the full symmetric group of degree mn, except in the cases $m = n = 2$, when we get a cyclic group of order 4, and $m = 2, n = 3$, when we get S_5. To show this, we can use the fact that the symmetric group on $\{1, 2, \ldots, n\}$ is generated by transpositions of the form $(k, k+1)$, where $1 \le k \le n-1$.

6.13 Stamp Rolls

We have two stamp rolls with unlimited supplies of 1-cent and 2-cent stamps (Figure 6.6).

Let $a(n)$ be the number of ways to make postage of n cents by taking strips of stamps from the two rolls. The order of the strips and the number of stamps per strip matter. For

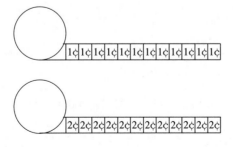

Figure 6.6. One-cent and two-cent stamp rolls.

example, $a(4) = 15$, as there are fifteen ways to make postage of four cents:

$$(1) + (1) + (1) + (1), \ (1 + 1) + (1) + (1), \ (1) + (1 + 1) + (1),$$

$$(1) + (1) + (1 + 1), \ (1 + 1) + (1 + 1), \ (1 + 1 + 1) + (1),$$

$$(1) + (1 + 1 + 1), \ (1 + 1 + 1 + 1), \ (2) + (1) + (1),$$

$$(1) + (2) + (1), \ (1) + (1) + (2), \ (2) + (1 + 1),$$

$$(1 + 1) + (2), \ (2) + (2), \ (2 + 2).$$

In this notation, the numbers within parentheses comprise a strip. For instance, $(2) + (1 + 1)$ means a single 2-cent stamp followed by a strip of two 1-cent stamps.

Find $a(100)$ and give an approximate value of $a(10^5)$, the number of ways to make postage of \$1000. You will probably want to use a computer.

We will find a recurrence relation for $\{a(n)\}$. Then we can calculate $a(100)$ and approximate $a(10^5)$. When solving a challenging mathematics problem, it's a good idea to look at specific cases. We can easily find $a(1) = 1$, $a(2) = 3$, $a(3) = 6$, and $a(4) = 15$ (as above). We can go a little further and find $a(5) = 33$ and $a(6) = 78$, but larger values are harder to produce. Can you guess a recurrence relation for $\{a(n)\}$ from what we know?

Since the last strip chosen consists of some number of 1-cent or 2-cent stamps, we have the recurrence relation

$$a(n) = a(n - 1) + a(n - 2) + a(n - 3) + a(n - 4) + \cdots$$
$$+ a(n - 2) + a(n - 4) + a(n - 6) + a(n - 8) + \cdots,$$

where we stop summing when the arguments become negative, and we define $a(0) = 1$. We could use this to calculate $a(100)$, but it would be better to find a recurrence relation that requires only a fixed number of previous terms. For $n \geq 3$, we have

$$a(n) = a(n - 1) + a(n - 2) + [a(n - 3) + a(n - 4) + \cdots]$$
$$+ a(n - 2) + [a(n - 4) + a(n - 6) + a(n - 8) + \cdots]$$
$$= a(n - 1) + 3a(n - 2).$$

Hence, $\{a(n)\}$ is given by a linear recurrence relation of order two:

$$a(0) = 1, \ a(1) = 1, \ a(2) = 3,$$
$$a(n) = a(n - 1) + 3a(n - 2), \quad n \geq 3.$$

With the help of a computer, we find that

$$a(100) = 8703381418732146559195732006487001751 \doteq 8.7 \times 10^{35}.$$

We can calculate $a(10^5)$ by computer using the recurrence relation and get an approximate answer:

$$a(10^5) \doteq 7.5 \times 10^{36224}.$$

Or we can obtain an exact formula for $a(n)$ and then approximate it. The standard way to solve a linear homogeneous recurrence relations of order two with constant coefficients is to assume it has a solution of the form

$$a(n) = \alpha r_1^n + \beta r_2^n, \quad n \geq 1,$$

where α and β are constants, and r_1 and r_2 are the distinct roots of the *characteristic equation*

$$x^2 - x - 3 = 0,$$

i.e.,

$$r_1 = \frac{1 - \sqrt{13}}{2}, \quad r_2 = \frac{1 + \sqrt{13}}{2}.$$

Using $a(1) = 1$ and $a(2) = 3$, we find

$$\alpha = \frac{1}{3} - \frac{2}{3\sqrt{13}} \quad \text{and} \quad \beta = \frac{1}{3} + \frac{2}{3\sqrt{13}},$$

thus obtaining

$$a(n) = \left(\frac{1}{3} - \frac{2}{3\sqrt{13}}\right) r_1^n + \left(\frac{1}{3} + \frac{2}{3\sqrt{13}}\right) r_2^n, \quad n \geq 1.$$

For large n, the exponential term r_2^n dominates. Hence

$$a(10^5) \doteq \left(\frac{1}{3} + \frac{2}{3\sqrt{13}}\right) r_2^{100000} \doteq 7.5 \times 10^{36224}.$$

We can also approximate $a(10^5)$ using a generating function. Define

$$f(x) = \sum_{n=0}^{\infty} a(n)x^n = a(0) + a(1)x + a(2)x^2 + a(3)x^3 + a(4)x^4 + \cdots$$

$$= 1 + x + 3x^2 + 6x^3 + 15x^4 + \cdots.$$

The recurrence relation for $\{a(n)\}$ yields

$$\sum_{n=0}^{\infty} a(n)x^n (1 - x - x^2 - x^3 - x^4 - \cdots - x^2 - x^4 - x^6 - x^8 - \cdots) = a(0) = 1.$$

From the formula for the sum of a geometric series, we obtain

$$f(x) = \frac{1}{1 - \frac{x}{1-x} - \frac{x^2}{1-x^2}} = \frac{1 - x^2}{1 - x - 3x^2}.$$

The generating function is a rational function and the coefficients in its denominator match those in the recurrence relation. This happens with the generating function of any linear recurrence relation with constant coefficients.

We can write the generating function as

$$f(x) = \frac{1}{3} + \frac{\alpha}{1 - r_1 x} + \frac{\beta}{1 - r_2 x}.$$

The geometric series with growth rate $r_2 x$ dominates, and we obtain the same approximation as before:

$$a(10^5) \doteq \beta r_2^{100000} \doteq 7.5 \times 10^{36224}.$$

6.14   Making a Million

How many ways can you make \$1 million using any number of pennies, nickels, dimes, quarters, one-dollar bills, five-dollar bills, ten-dollar bills, twenty-dollar bills, fifty-dollar bills, and hundred-dollar bills? You will need a computer for this problem.

The number of pennies used in making a million dollars must be a multiple of 5. Thus, we may think of 5 cents (a nickel or five pennies) as the smallest unit of currency, all other units being multiples of 5. Let c_n be the number of ways to make $5n$ cents. The generating function for the sequence $\{c_n\}$ is

$$
\begin{aligned}
c(x) =\,&1 + c_1 x + c_2 x^2 + c_3 x^3 + \cdots \\
=\,&\frac{1}{1-x}\frac{1}{1-x}\frac{1}{1-x^2}\frac{1}{1-x^5} \\
&\cdot \frac{1}{1-x^{20}}\frac{1}{1-x^{100}}\frac{1}{1-x^{200}}\frac{1}{1-x^{400}}\frac{1}{1-x^{1000}}\frac{1}{1-x^{2000}}.
\end{aligned}
$$

The terms in the middle row above correspond to the contributions from the coins. The terms in the bottom row represent contributions from the paper money. To see how the factors in the generating function work, consider the contribution to the generating function from the term

$$
\frac{1}{1-x^2} = 1 + x^2 + x^{2\cdot 2} + x^{3\cdot 2} + x^{4\cdot 2} + \cdots .
$$

A selection of, say, $x^{3\cdot 2}$ corresponds to a selection of three dimes (a dime is two of our basic units). When we multiply all selections together and combine coefficients of like powers, we obtain the generating function for $\{c_n\}$.

Since \$1 million is 20,000,000 of the 5-cent units, our job is to find $c_{20,000,000}$, the coefficient of $x^{20,000,000}$ in $c(x)$. We will write the generating function in a form that makes this easier.

In the denominator of $c(x)$, all the powers of x divide the largest power, 2000. Accordingly, we rewrite each factor in the denominator as $(1 - x^{2000})$ with a compensating factor, a geometric series, in the numerator. The new numerator is

$$
(1 + x + \cdots + x^{1999})^2 (1 + x^2 + x^4 + \cdots + x^{1998})(1 + x^5 + \cdots + x^{1995})
$$
$$
\cdot (1 + x^{20} + \cdots + x^{1980})(1 + x^{100} + \cdots + x^{1900})(1 + x^{200} + \cdots + x^{1800})
$$
$$
\cdot (1 + x^{400} + \cdots + x^{1600})(1 + x^{1000}).
$$

A computer algebra system can quickly multiply out the new numerator. The new denominator is $(1 - x^{2000})^{10}$, and we can expand its reciprocal as a binomial series:

$$
(1 - x^{2000})^{-10} = \sum_{k=0}^{\infty} \binom{k+9}{9} x^{2000k}.
$$

We complete the calculation by multiplying this binomial series by appropriate terms from the numerator to obtain the coefficient of $x^{20,000,000}$. The numerator is a polynomial, say

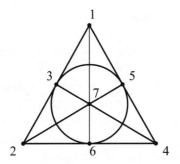

Figure 6.7. A projective plane of order 2.

p, of degree 16271, but the only powers of x that matter are multiples of 2000; the corresponding coefficients are

$$p_0 = 1$$

$$p_{2000} = 48820947949$$

$$p_{4000} = 3864246773424$$

$$p_{6000} = 34961841233371$$

$$p_{8000} = 73423441820500$$

$$p_{10000} = 41833663537539$$

$$p_{12000} = 5760824622356$$

$$p_{14000} = 107159417621$$

$$p_{16000} = 1647239.$$

We calculate, again with the help of a computer,

$$c_{20,000,000} = \sum_{j=0}^{8} \binom{10000 - j + 9}{9} p_{2000j}$$

$$= 441,287,168,799,272,062,712,629,114,612,633,953,025,220,001$$

$$\doteq 4.4 \times 10^{44}.$$

6.15 Coloring a Projective Plane

A projective plane of order three is shown on page 17. We will consider a projective plane of order two, as shown in Figure 6.7. It has seven points and seven lines. Each line contains three points, each point lies on three lines, every two points determine a unique line, and every two lines intersect in a unique point.

If we color the points with different colors, how many different colorings do we obtain? The points and lines of a finite geometry may be moved around as long as the incidences between points and lines remain unchanged. For instance, the points 1, 2, and 3 must be collinear no matter how we draw the plane.

We will show that there are 30 different colorings. The important thing is the number of symmetries of the plane that can occur when the points are moved. Since all seven points are equivalent, we may move any point to occupy the position of any other point. So there are seven choices for where to move a point. Once that choice is made, there are six choices for where to move another point. However, once these decisions are made, the third point collinear with the first two must stay on the line determined by them. There are four remaining points, and any of them can be moved to any of the four remaining positions. However, this choice, together with the earlier choices, determines the positions of all the remaining points. Altogether, there are $7 \cdot 6 \cdot 4 = 168$ choices, and this is the number of symmetries of the projective plane.

Without symmetries, we would have $7! = 5040$ different colorings. This multiply-counts colorings that can be obtained from each other by a symmetry. Since there are 168 symmetries, there are only $5040/168 = 30$ different colorings.

Recalling A Group of Operations from Chapter 5, you may notice that the formula for the number of symmetries of the projective plane, $7 \cdot 6 \cdot 4$, gives the number of invertible 3×3 binary matrices. The multiplicative group of the matrices is isomorphic to the symmetry group of the projective plane. We can specify the isomorphism by labeling the seven points of the projective plane with the seven nonzero binary vectors of length three. Each matrix acts by multiplication on the set of points. We must do the labeling so that the vectors corresponding to three points on a line sum to 0, as collinearity is preserved by matrix multiplication.

7

Creative Problems

In mathematics, you understand what you build up.

—FAN CHUNG

It is easy to formulate new mathematical problems. One only needs an inquiring mind. The problems in this chapter are partially or completely unsolved, so there is much to work on!

7.1 Two-Dimensional Gobbling Algorithm

Choose a positive integer, say, 20. Now choose a random integer between 1 and 20, say, 9. Subtract: $20 - 9 = 11$. Next, choose a random integer between 1 and 11, say, 7. Subtract: $11 - 7 = 4$. Choose a random integer between 1 and 4, say, 3. Subtract: $4 - 3 = 1$. Now we must choose the integer 1, and we subtract: $1 - 1 = 0$. Since we have obtained 0, we stop. We did four subtractions. Starting with 20, how many subtractions are expected?

We can show, using a recurrence relation, that starting with a positive integer n, the expected number of subtractions is the *harmonic number*

$$H_n = 1 + \frac{1}{2} + \frac{1}{3} + \cdots + \frac{1}{n}.$$

What if we start with a pair of positive integers, say, $(10, 6)$? Choose a random integer between 1 and 10, say, 5, and a random integer between 1 and 6, say, 2. Subtract: $(10, 6) - (5, 2) = (5, 4)$. Repeat, choosing the ordered pair $(2, 3)$. Subtract: $(5, 4) - (2, 3) = (3, 1)$. Repeat, choosing $(2, 1)$. Subtract: $(3, 1) - (2, 1) = (1, 0)$. Since one of the numbers is 0, we stop. There were three subtractions. How many subtractions do we expect?

Let $e(m, n)$ be the expected number of subtractions, starting with a pair of positive integers (m, n). It's easy to write a recurrence formula for it:

$$e(m, 1) = 1, \quad m \geq 1;$$

$$e(1, n) = 1, \quad n \geq 1;$$

$$e(m, n) = 1 + \frac{1}{mn} \sum_{j=1}^{m-1} \sum_{k=1}^{n-1} e(m, n), \quad m, n > 1.$$

The recurrence formula produces a table of values of $e(m, n)$.

		n				
		1	2	3	4	5
m	1	1	1	1	1	1
	2	1	5/4	4/3	11/8	7/5
	3	1	4/3	53/36	223/144	115/72
	4	1	11/8	223/144	475/288	549/320
	5	1	7/5	115/72	549/320	4309/2300

Do you see a pattern?

7.2 Nonattacking Queens Game

A chess Queen attacks all squares on its row, column, and diagonals. A Queen's range on an 8×8 board is shown in Figure 7.1. Let n be a positive integer. Two players play a game in which they alternately place Queens on an $n \times n$ board so that each new Queen is out of range of the others. The last player able to place a Queen on the board is the winner.

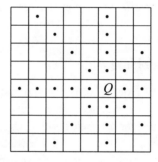

Figure 7.1. A Queen's range on an 8×8 board.

Given best possible play by both players, who should win this game? If n is odd, then the first player has a winning strategy: place the first Queen in the center of the board and after a second player move, take the square symmetric with respect to the center. Every time the second player has an available move, so does the first player, so the second player will run out of moves first. For n even, the outcome of the game is in general unknown. This problem was introduced in [18]. You can show by considering all cases that the first player wins for $n = 2, 4, 6,$ and 8. Hassan Noon and Glen van Brummelen [38] showed that the second player wins for $n = 10$. Can you determine who wins on a 12×12 board?

7.3 Lucas Numbers Mod m

The *Lucas numbers* are defined by

$$L_0 = 2, \; L_1 = 1, \quad L_n = L_{n-1} + L_{n-2}, \quad n \geq 2.$$

The sequence $\{L_n\}$ satisfies the same recurrence relation as the Fibonacci sequence but has different initial values.

Given $m \geq 2$, when does the range of the sequence $\{L_n \mod m\}$ consist of a complete residue system modulo m? The corresponding question for Alcuin's sequence was addressed in Chapter 5. Although we don't have a formula for the period of $\{L_n \mod m\}$, we know by the pigeonhole principle that it is at most m^2.

Computer explorations give rise to a conjecture.

Conjecture. The sequence $\{L_n \mod m\}$ takes all values modulo m if and only if m is one of

$$2,\ 4,\ 6,\ 7,\ 14,\ 3^k,\quad k \geq 1.$$

For example, the sequence $\{L_n \mod 6\}$ is

$$2,\ 1,\ 3,\ 4,\ 1,\ 5,\ 0,\ \ldots,$$

and we obtain all the residues modulo 6. The sequence $\{L_n \mod 5\}$ is

$$2,\ 1,\ 3,\ 4,\ 2,\ 1,\ \ldots,$$

and since it repeats we never obtain the residue 0.

Can you prove the conjecture?

Stephen A. Burr solved the corresponding problem for the Fibonacci sequence [11]. The sequence $\{F_n \mod m\}$ contains all residues modulo m if and only if m is one of

$$5^j,\ 2 \cdot 5^j,\ 4 \cdot 5^j,\ 3^k \cdot 5^j,\ 6 \cdot 5^j,\ 7 \cdot 5^j,\ 14 \cdot 5^j,\quad j \geq 0,\ k \geq 1.$$

7.4 Exact Colorings of Graphs

A graph of the kind encountered in graph theory is a set of vertices and a set of edges joining pairs of vertices. A *complete graph* is a graph in which every two vertices are joined by an edge.

There are many problems about colorings of graphs. In this problem, we are concerned with coloring the edges of a graph using a set of colors. An *exact c-coloring* of a graph is an assignment of one color chosen from c colors, to each edge of the graph such that each color is used at least once.

The following exact coloring problem for infinite graphs is unsolved. For $1 \leq m \leq c$, let $P(c, m)$ be the statement that every exact c-coloring of the edges of a countably infinite complete graph yields an exactly m-colored countably infinite complete subgraph. For what values of c and m is $P(c, m)$ true? The case $m = 1$ is a famous theorem of combinatorics known as Ramsey's theorem.

Ramsey's Theorem. *If the edges of the complete infinite graph on a countable infinity of vertices are colored using finitely many colors, then there exists a complete subgraph on infinitely many vertices all of whose edges are the same color.*

A corollary of Ramsey's theorem is that $P(c, m)$ is true when $m = 2$. If $c = m$, then $P(c, m)$ is trivially true (take the subgraph to be the given graph). These may be the only values of c and m for which $P(c, m)$ is true.

Conjecture. The statement $P(c, m)$ is true if and only if $m = 1$, $m = 2$, or $c = m$.

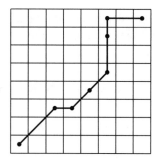

Figure 7.2. A Queen path.

As an example of how we can disprove $P(c, m)$, consider the case $P(11, 5)$. We exhibit an exact 11-coloring of the edges of the complete infinite graph on countably many vertices so that there is no exactly 5-colored complete infinite subgraph. Color each edge of a subgraph K_5 using a different color. This requires $\binom{5}{2} = 10$ colors. Color every other edge in the infinite graph using the 11th color. Every infinite subgraph is exactly $\binom{k}{2} + 1$ colored for some k. Since 5 is not a number of this form, $P(11, 5)$ is false.

Alan Stacey and Peter Weidl [49] proved that $P(c, m)$ is false for each fixed $m \geq 3$ and c sufficiently large. Can you prove the conjecture in its entirety?

7.5   Queen Paths

A chess Queen can move any number of squares horizontally, vertically, or diagonally in one step. Figure 7.2 shows a sample Queen path from the lower-left corner of the board to the upper-right corner.

Let us extend the board infinitely to the right and upward. Denote the squares by ordered pairs of nonnegative integers, with the lower-left corner square labeled $(0, 0)$.

How many lattice paths can the Queen take from $(0, 0)$ to (m, n), where m and n are nonnegative integers? Let $q(m, n)$ be the number of paths from $(0, 0)$ to (m, n) such that at each step the Queen moves up, right, or up-right. In the table, we calculate each entry by adding all the entries to the left of, below, and diagonally left-below the entry. For example, $q(3, 2) = 2 + 7 + 22 + 4 + 17 + 1 + 7 = 60$. The reason this works is that the Queen has to arrive at the given square from one of these squares.

⋮	⋮	⋮	⋮	⋮	⋮	⋮	⋮	
64	464	2392	10305	39625	140658	470233	1499858	...
32	208	990	3985	14430	48519	154352	470233	...
16	92	401	1498	5079	16098	48519	140658	...
8	40	158	543	1712	5079	14430	39625	...
4	17	60	188	543	1498	3985	10305	...
2	7	22	60	158	401	990	2392	...
1	3	7	17	40	92	208	464	...
1	1	2	4	8	16	32	64	...

We see from the table that the number of Queen paths from the lower-left corner to the upper-right corner of the board is $q(7, 7) = 1499858$.

The recurrence relation for the two-variable sequence requires arbitrarily many prior values. Let's find a recurrence relation for the number of Queen paths that requires a fixed number of prior values. We use the generating function method that we saw in the solution to Stamp Rolls in Chapter 6.

We represent the Queen's basic steps by the indeterminates x, y, and xy. From the recurrence relation that we have already found, we have

$$\sum_{m=0}^{\infty} \sum_{n=0}^{\infty} q(m,n) x^m y^n (1 - x - x^2 - \cdots - y - y^2 - \cdots - (xy) - (xy)^2 - \cdots)$$

$$= q(0,0) = 1.$$

Thus, the generating function is

$$\sum_{m=0}^{\infty} \sum_{n=0}^{\infty} q(m,n) x^m y^n = \frac{1}{1 - \frac{x}{1-x} - \frac{y}{1-y} - \frac{xy}{1-xy}}$$

$$= \frac{1 - x - y + x^2 y + xy^2 - x^2 y^2}{1 - 2x - 2y + xy + 3x^2 y + 3xy^2 - 4x^2 y^2}.$$

Looking at the denominator of the generating function, we can read off a recurrence formula for the number of Queen paths:

$$q(0,0) = 1, \ q(0,1) = 1, \ q(0,2) = 2,$$

$$q(1,0) = 1, \ q(1,1) = 3, \ q(1,2) = 7,$$

$$q(2,0) = 2, \ q(2,1) = 7, \ q(2,2) = 22;$$

$$q(m,n) = 2q(m-1,n) + 2q(m,n-1) - q(m-1,n-1) - 3q(m-2,n-1)$$

$$- 3q(m-1,n-2) + 4q(m-2,n-2), \quad m \geq 3 \text{ or } n \geq 3.$$

We set $q(m,n) = 0$ for m or n negative.

The *diagonal sequence* for Queen paths, $\{q_n = q(n,n)\}$, is

$$1, 3, 22, 188, 1712, 16098, 154352, 1499858, 14717692, 145509218, \ldots.$$

Its generating function is

$$\frac{(x-1)}{(3x-2)} \left[1 + \frac{1-x}{\sqrt{1-12x+16x^2}} \right].$$

From the generating function we obtain a recurrence formula:

$$q_0 = 1, \ q_1 = 3, \ q_2 = 22, \ q_3 = 188;$$

$$q_n = ((29n-18)q_{n-1} + (-95n+143)q_{n-2}$$

$$+ (116n-302)q_{n-3} + (-48n+192)q_{n-4})/(2n), \quad n \geq 4.$$

The quantity under the square root sign can be factored as

$$1 - 12x + 16x^2 = (1 - r_1 x)(1 - r_2 x),$$

where $r_1, r_2 = 6 \pm 2\sqrt{5}$. It can be shown that

$$q_n \sim c\, r_1^n / \sqrt{\pi n},$$

where $c = \sqrt{10(3\sqrt{5} - 5)}/8$.

The problem of counting Queen paths can be generalized to higher dimensions.

A Queen path from $(0,0,0)$ proceeds in steps that are positive integer multiples of $(1,0,0)$, $(0,1,0)$, or $(0,0,1)$.

A Queen path from $(0,0,\ldots,0)$ to $(a_1, a_2, \ldots, a_d)$ is equivalent to a Wythoff's Nim game that starts with d piles of stones of sizes $a_1, a_2, \ldots, a_d$. In the game, two players alternately remove the same number of stones from any of the piles. The game is over when the last stone is removed. Our formulas count the number of possible games.

The number of Queen paths to a main diagonal point (n, n, n) has been recently conjectured by Alin Bostan (leading a team) to satisfy a linear recurrence relation of order 14 with polynomial coefficients of degree 52.

What is a recurrence relation for the number of Queen paths to a diagonal point in dimension $d \geq 3$?

7.6 Transversal Achievement Game

Recall from Transversal of Primes in Chapter 1 that a transversal of an $n \times n$ array is a set of n cells with no two in the same row or column. We can define a two-player game based on transversals. Two players, Oh and Ex, alternately choose unoccupied cells from an $n \times n$ array. They write their symbols, O and X, in the chosen cells. The first player, if any, to occupy a collection of n cells constituting a transversal is the winner. (The player may occupy other cells, too.) If there is a winner, then the winner must be the first player, Oh. The reason is that if the second player, Ex, had a winning strategy, then the first player could adopt it and win one step earlier. A blocking strategy for Ex might be to try to occupy a complete row or column. This would prevent Oh from occupying a transversal. However, Oh may prevent Ex from occupying a complete row or column. For what values of n does Oh win this game?

7.7 Binary Matrix Game

A two-dimensional version of van der Waerden's theorem, called Gallai's theorem, named after Tibor Gallai (1912–1992), guarantees that in any coloring of the elements of an infinite square grid with two colors, there must exist four cells, all colored the same, lying at the vertices of a square with horizontal and vertical sides. A finite version of Gallai's theorem says that there exists a positive integer n such that given any two-coloring of the cells of an $n \times n$ grid, there exist some four cells all the same color, lying at the vertices of a square with horizontal and vertical sides. We can think of the grid as a matrix and take the colors to be 0 and 1, so that we have a binary matrix.

Say that a *constant sub-square* is a set of four equal entries of a binary matrix at the vertices of a square with horizontal and vertical sides. A long-standing problem was to find the least value of n that forces the existence of a constant sub-square. It was solved in

Figure 7.3. A 14×14 binary matrix without a constant sub-square.

Figure 7.4. A pattern for a $13 \times \infty$ binary matrix without a constant sub-square.

2009 by Roland Bacher and Shalom Eliahou [5], who proved that $n = 15$. Furthermore, they showed that every 14×15 binary matrix must have four such entries, and there exist 14×14 and $13 \times \infty$ binary matrices that don't.

Figure 7.3 shows a 14×14 binary matrix with no constant sub-square. Figure 7.4 shows the pattern for a $13 \times \infty$ binary matrix with no constant sub-square. Representing 0 by a blank and 1 by a filled square gives the figures, especially the second one, an Escher-like quality.

We can create a game related to the Bacher–Eliahou result. Suppose that two players, Oh and Ex, alternately place their symbols, O and X, in unoccupied cells of an $n \times n$ grid. The first player, if any, to mark four cells of a constant sub-square is the winner. As in the transversal achievement game, if there is a winner with best possible play, then it is Oh. The proof is by contradiction. If Ex had a winning strategy, then Oh could simply adopt it and get there first. By the Bacher–Eliahou result, Oh has a winning strategy if $n \geq 15$. But perhaps Oh can force a win on a smaller playing board. What is the least value of n for which Oh can always win?

If instead of a binary matrix we have a trivalued matrix with entries 0, 1, or 2, then the minimum size of the matrix that guarantees the existence of a constant sub-square is unknown. Can you find it?

A

Harmonious Foundations

> Mathematics is a more powerful instrument of knowledge than any other
> that has been bequeathed to us by human agency.
>
> —René Descartes (1596–1650)

Mathematical definitions appear inevitable, as if they exist independently of human thought. The appearance of inevitability prompts the question of whether mathematics is discovered or invented. We can't answer that question, but we note that someone had to think of the definitions that we now take for granted. This results from a historical process of formulating problems, looking for solutions, and creating the best mathematics for the given situations. In this appendix, we give background information on the mathematical concepts in the book. As a utilitarian fork or a chair can be beautiful, everyday mathematical constructs are also beautiful. Simple definitions can give rise to surprising phenomena. A good reference on mathematical foundations is [48].

A.1   Sets

Sets provide the building blocks for many mathematical definitions. The modern notion of sets was introduced by Georg Cantor (1845–1918). However, Cantor's set theory admitted some paradoxes, the most famous of which is Russell's paradox. It concerns the set S of all sets that are not members of themselves. If S is a member of itself, then by definition S is not a member of itself. But if S is not a member of itself, then by definition S is a member of itself. There is a contradiction either way. Set theory was put on a firm foundation by Ernst Zermelo (1871–1953) and Abraham Fraenkel (1891–1965). Their system, together with the Axiom of Choice, is called *ZFC set theory*. Russell's paradox has found its way into rigorous mathematics via results in mathematical logic such as those due to Kurt Gödel (1906–1978). Gödel's Incompleteness Theorem asserts that in any consistent mathematical system, that is, one in which false statements are not provable, and rich enough to contain the integers, there exists a statement $\mathfrak{G}$ that is true but not provable within the system. That is, the system is incomplete.

Statement $\mathfrak{G}$: Statement $\mathfrak{G}$ is not provable within the system.

Consider whether $\mathfrak{G}$ is true or false. If it is false, then it is provable within the system. But this would mean that a false statement is provable, which is impossible in a consistent system. Hence $\mathfrak{G}$ is a true statement. Since $\mathfrak{G}$ says that $\mathfrak{G}$ is not provable within the system, this must be the case. So we have a statement, $\mathfrak{G}$, that is true but not provable within the system. Thus the system is incomplete.

A virtue of set theory is that we can define many important mathematical objects in terms of sets. For a clear introduction to Gödel's theory, see [35]. For an advanced discussion of the way set theory is used in mathematical applications, see [12].

A *set* is a collection of *elements*. We sometimes define a set by listing its elements. For example,

$$A = \{1, 2, 3, 4, 5, 6, 7, 8, 9, 10\}$$

has for its elements the integers 1 and 10 inclusive.

We may also define a set by a rule that its elements satisfy. Thus A may be written as

$$A = \{x : x \text{ is an integer between 1 and 10 (inclusive)}\}.$$

We write $x \in S$ to indicate that x is an element of S. Thus $3 \in A$. If two sets A and B have the same elements, then A and B are *equal* and we write $A = B$.

We say that A is a *subset* of B, and write $A \subseteq B$, if every element of A is an element of B. If $A \subseteq B$ and $B \subseteq A$, then by definition $A = B$.

The *empty set*, denoted by $\emptyset$, is the set with no elements.

Some sets of numbers have special names:

$\mathbf{N} =$ the set of *natural numbers* $\{1, 2, 3, \ldots\}$

$\mathbf{Z} =$ the set of *integers* (positive, negative, and 0)

$\mathbf{Q} =$ the set of *rational numbers*

$\mathbf{R} =$ the set of *real numbers*

$\mathbf{C} =$ the set of *complex numbers*.

The *cardinality* of a set A, denoted by $|A|$, is the number of elements in A. For example, $|\{1, 3, 5, 7, 9\}| = 5$. If A has finitely many elements, we say that A is *finite*. If A is not finite, then A is *infinite*. Two sets are said to have the cardinality if there is a bijection between them.

The *union* of two sets A and B, written $A \cup B$, is the set of elements in A or B or both. The *intersection* of A and B, written $A \cap B$, is the set of elements in both A and B.

Sets A and B are *disjoint* if $A \cap B = \emptyset$. If A and B are disjoint, then $|A \cup B| = |A| + |B|$. A collection of sets $\mathcal{C}$ is *pairwise-disjoint* if every pair of members of $\mathcal{C}$ are disjoint.

The *difference* of A and B, written $A - B$, is the set of elements in A but not in B. If $B \subseteq A$, and the set A is clear from context, we call $A - B$ the *complement* of B, and denote it by $\overline{B}$.

The *power set* of A, denoted $\mathcal{P}(A)$, is the collection of all subsets of A, including the empty set. If the cardinality of A is n, then $|\mathcal{P}(A)| = 2^n$.

It is always the case that the cardinality of $\mathcal{P}(A)$ is greater than the cardinality of A (even if A is an infinite set). The reason is that there is no onto function from A to $\mathcal{P}(A)$. For suppose that f is a function from A to $\mathcal{P}(A)$. Let $X = \{x \in A : x \notin f(x)\}$. Given any a in A, if $a \in X$, then $a \notin f(a)$, and hence $f(a) \neq X$; and if $a \notin X$, then $a \in f(a)$, and hence $f(a) \neq X$. Therefore, X is not in the range of f and we conclude that f is not an onto function. As a consequence of this result, there is no largest infinite set.

A.2   Relations

The *Cartesian product* $A \times B$ of two sets A and B is the collection of all ordered pairs (a, b) with $a \in A$ and $b \in B$. If A and B are finite, then $|A \times B| = |A||B|$.

A *relation* R on a set X is a subset of $X \times X$. If $(a, b) \in R$, then a is *related* to b.

Here are two relations on the set $\mathbf{Z}$ of integers:

$$R_1 = \{(a, b) : a, b \in \mathbf{Z} \text{ and } a - b \text{ is divisible by } 3\}$$

and

$$R_2 = \{(a, b) : a, b \in \mathbf{Z} \text{ and } a \text{ is divisible by } b\}.$$

A relation R on X is *reflexive* if $(a, a) \in R$ for all $a \in X$; *symmetric* if $(b, a) \in R$ whenever $(a, b) \in R$; *antisymmetric* if $(a, b) \in R$ and $(b, a) \in R$ imply that $a = b$; and *transitive* if $(a, b) \in R$ and $(b, c) \in R$ imply that $(a, c) \in R$.

An *equivalence relation* is a relation that is reflexive, symmetric, and transitive. The relation R_1 is an equivalence relation. Given an element $x \in X$, the set of elements related to x, that is, the set of $y \in X$ such that $(x, y) \in R$, is called the *equivalence class* of x, and is denoted $[x]$. For example, in R_1 we have $[0] = \{\ldots, -6, -3, 0, 3, 6, \ldots\}$.

In general, for $m \geq 2$, we define the *congruence relation* $a \equiv b$ (modulo m) if and only if $a - b$ is divisible by m. This is an equivalence relation on $\mathbf{Z}$.

A *partial order* is a relation that is reflexive, antisymmetric, and transitive. The relation R_2 is a partial order.

A.3   Functions

The concept of a function is used so often that it may be difficult to understand why there was ever any ambiguity about the definition. Is a function a machine that takes an input and gives an output? Is it a curve that you can graph? It may seem strange that the modern definition of function wasn't worked out until the twentieth century. Before then, a function could conceivably return more than one value for a given input, so it was more general than the functions of today which return only one value. On the other hand, until the 1800s a function was a mapping that could be constructed from a family of well known functions such as sine, cosine, and exponential functions.

A *function* is a set of ordered pairs (x, y), where x is an element of a set X and y is an element of a set Y, where it is possible that $X = Y$. Every element of X occurs as the first element of an ordered pair, and for each x in X there is a unique corresponding y in Y. Less formally, a function is a rule that assigns to each element in X a unique element in Y.

There is no requirement that a function can be graphed by a continuous curve. An example of a *pathological function*, nowhere continuous, was given by Peter Gustav Lejeune Dirichlet (1805–1859). It is a function from the set of real numbers to the set $\{0, 1\}$. The value of the function is 1 at each rational number and 0 at each irrational number. You can't graph the function because it oscillates too wildly between 0 and 1. However, it is well-defined.

A *function* f from A to B, written $f: A \to B$, is a subset of $A \times B$ such that for each $a \in A$ there exists a unique $b \in B$ with $(a, b) \in f$. We say that f *maps* a to b, and we write $f(a) = b$ or $f: a \mapsto b$. We call b the *image* of a. The *domain* of f is A. The *codomain* of f is B. The *range* of f is the set of $b \in B$ for which there exists $a \in A$ such that $f(a) = b$.

For example, the function

$$f: \{1, 2, 3, 4\} \longrightarrow \{1, 2, 3, 4, 5, 6, 7, 8\}$$
$$x \mapsto 2x$$

maps each element $x \in \{1, 2, 3, 4\}$ to its double in $\{1, 2, 3, 4, 5, 6, 7, 8\}$. The range of f is $\{2, 4, 6, 8\}$.

The *identity function* on A is the function $f: A \to A$ defined by $f(x) = x$.

A function $f: A \to B$ is *one-to-one* if no two elements of A are mapped to the same element of B. The function f is *onto* if each $b \in B$ is the image of some $a \in A$. Equivalently, f is onto if the range of f equals B. If f is one-to-one and onto, then f is a *bijection*. If f is a bijection, then the *inverse* of f, denoted f^{-1}, is a function from B to A where $f^{-1}(b) = a$ if and only if $f(a) = b$.

The following theorem is often useful.

Theorem. *Suppose that X and Y are finite sets of the same cardinality. Then a function from X to Y is one-to-one if and only if it is onto.*

Given functions $f: A \to B$ and $g: B \to C$, the *composition* of f and g is the function from A to C defined by $x \mapsto g(f(x))$. If $f: A \to A$ is a bijection, then the composition of f and f^{-1} is the identity function on A.

A.4   Groups

The mathematical term *group* was first used by Évariste Galois (1811–1832) in the study of the solvability of polynomial equations. Other mathematicians, such as Arthur Cayley (1821–1895) and Augustin-Louis Cauchy (1789–1857), used essentially the same idea in the study of permutations. Eventually, these ideas, and others involving number theory and geometry, were synthesized into the modern definition of an abstract group. A good primer on group theory is [25].

A *group* G is a nonempty set together with a binary operation $*$ such that:

For all $x, y \in G$, we have $x * y \in G$ (closure).

For all $x, y, z \in G$, we have $x * (y * z) = (x * y) * z$ (associative law).

There exists an element $e \in G$ with the property that, for all $x \in G$, we have $x * e = e * x = x$.

For every $x \in G$, there exists an element $x^{-1} \in G$ with the property that $x * x^{-1} = x^{-1} * x = e$.

The element e is called the *identity* of G. The element x^{-1} is called the *inverse* of x. The identity element of a group is unique and the inverse x^{-1} of each element x is unique.

Examples of groups:

The set of integers $\mathbf{Z}$ is a group with respect to addition.

The set $\mathbf{R} - \{0\}$ of nonzero real numbers is a group with respect to multiplication.

In writing group elements, we usually suppress the group operation sign, denoting $x * y$ by xy. We abbreviate xx by x^2, $x^{-1}x^{-1}$ by x^{-2}, etc. For all $x \in G$, we set $x^0 = e$.

A *finite group* is a group with a finite number of elements. The *order* of a finite group is the number of elements in it.

The *cyclic group* $\mathbf{Z}_n$, of order n, is the set $\{0, \ldots, n - 1\}$ with the operation of addition modulo n.

If p is prime, then the nonzero residues modulo p form a cyclic group of order $p-1$, with multiplication modulo p. In general, for $n \geq 2$, the set of numbers m such that $1 \leq m < n$ and $\gcd(m, n) = 1$ form a group of order $\phi(n)$, with multiplication modulo n. The group is denoted $\mathbf{Z}_n^*$. For example, $\mathbf{Z}_{10}^* = \{1, 3, 7, 9\}$, under multiplication modulo 10.

A group G is *abelian* if $xy = yx$ for all $x, y \in G$. Otherwise, G is *nonabelian*. For example, the group $\mathbf{Z}$ is abelian.

The *order* of an element $x \in G$ is the least positive integer n for which $x^n = e$. If there is no such integer, then x has *infinite order*. For example, in $\mathbf{Z}_4$, the elements 0, 1, 2, 3 have orders 1, 4, 2, 4, respectively.

The *symmetric group* S_n consists of the $n!$ permutations of an n-element set, e.g., $\{1, 2, 3, \ldots, n\}$. The group operation $*$ is the composition of permutations (performing one permutation followed by the other permutation). The elements of S_n are conveniently written in *cycle notation*. Thus

$$(1, 2, 3)(4, 8)(3, 6, 7)(5)(9)(10)$$

is the element of S_{10} that maps 1 to 2 to 3 to 1, transposes 4 and 8, maps 3 to 6 to 7 to 3, and fixes 5, 9, and 10. To multiply two permutations together, find the result of the composition of the two bijections (reading left to right). For example,

$$(1, 2, 3)(4, 5) * (1, 2, 3, 4, 5) = (1, 3, 2, 4)(5).$$

Since $(1, 2)(1, 3) \neq (1, 3)(1, 2)$, the symmetric group S_n is nonabelian for $n \geq 3$.

Two groups G_1 and G_2 are *isomorphic* if there is a bijection (called an *isomorphism*) $\varphi: G_1 \to G_2$ that preserves multiplication: $\varphi(gh) = \varphi(g)\varphi(h)$, for all $g, h \in G_1$. For example, $\mathbf{Z}_{10}^*$ is isomorphic to $\mathbf{Z}_4$. Can you find an isomorphism?

Suppose that G_1 and G_2 are two groups. The *product* of G_1 and G_2, denoted $G_1 \times G_2$, is the set of ordered pairs $\{(g_1, g_2): g_1 \in G_1, g_2 \in G_2\}$ subject to the multiplication rule $(g_1, g_2) * (g_1', g_2') = (g_1 g_1', g_2 g_2')$.

The product $\mathbf{Z}_2 \times \mathbf{Z}_2$ is a four-element group. It is not isomorphic to $\mathbf{Z}_4$, for $\mathbf{Z}_2 \times \mathbf{Z}_2$ has three elements of order 2 while $\mathbf{Z}_4$ has only one. The group $\mathbf{Z}_2 \times \mathbf{Z}_3$ is isomorphic to $\mathbf{Z}_6$. Can you find an isomorphism?

A subset H of G is a *subgroup* of G if H is a group with respect to the group operation of G. For example, the two-element group $\{(1, 2)(3), (1)(2)(3)\}$ is a subgroup of the six-element group S_3.

The symmetric group S_n is especially important because every finite group is isomorphic to a subgroup of some S_n.

Theorem. *If G is a finite group of order n, then G is isomorphic to a subgroup of S_n.*

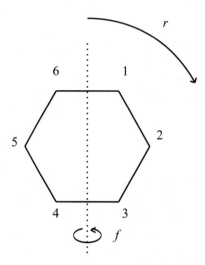

Figure A.1. Generators of the dihedral group D_6.

Here is a proof. For each element $g \in G$, we define a function $f_g \colon G \to G$ by the rule $f_g(a) = ag$ (right multiplication by g). Because f_g has an inverse, $f_{g^{-1}}$, it is a bijection. We check: $f_g(f_{g^{-1}}(a)) = ag^{-1}g = a$ and $f_{g^{-1}}(f_g(a)) = agg^{-1} = a$. Since f_g is a permutation of the n-element set G, we can define a function $\varphi \colon G \to S_n$ by $\varphi(g) = f_g$. We claim that φ is an isomorphism between G and the range of φ. First, we check that φ preserves multiplication: $\varphi(gh)(a) = f_{gh}(a) = a(gh) = (ag)h = f_h(f_g(a)) = (f_g f_h)(a) = (\varphi(g)\varphi(h))(a)$. Second, we check that φ is one-to-one: If $\varphi(g)(a) = \varphi(h)(a)$, then $f_g(a) = f_h(a)$, which implies that $ag = ah$, and $g = h$.

The *dihedral group* D_n, of order $2n$, consists of the set of symmetries of a regular convex n-gon. If we number the vertices of the n-gon $1, \ldots, n$, then we see that D_n is a subgroup of S_n. The subgroup is generated by two permutations: the *rotation* $r = (1, 2, 3, \ldots, n)$ and a *flip* f along an axis of symmetry of the n-gon. If n is odd, we take the flip to be

$$f = (n)(1, n - 1)(2, n - 2) \ldots ((n - 1)/2, (n + 1)/2).$$

If n is even, we take

$$f = (1, n)(2, n - 1) \ldots (n/2, n/2 + 1).$$

See Figure A.1 for a depiction of D_6, the group of symmetries of a regular convex hexagon.

Every element of D_n can be written in the form $r^\alpha f^\beta$, where $\alpha \in \{0, 1, 2, \ldots, n - 1\}$ and $\beta \in \{0, 1\}$. Elements are multiplied using the rules $r^n = e$, $f^2 = e$, and $rf = fr^{-1}$. We say that D_n has the *presentation*

$$\langle r, f \colon r^n = e, \ f^2 = e, \ rf = fr^{-1} \rangle.$$

For an explanation of the theory of group presentations, consult [27].

We have noted that every element of S_n can be expressed as a product of cycles. A cycle of length one is called a *fixed point* and a cycle of length two is called a *transposition*.

Cycles of length greater than two can be written as products of transpositions. For example, $(1, 2, 3) = (1, 2)(1, 3)$. A permutation may be written as a product of fixed points and transpositions in more than one way. The number of transpositions is always even or always odd. A permutation is accordingly called an *even permutation* or an *odd permutation*. Of the $n!$ permutations in S_n, half are even and half are odd. This follows from the observation that $f(\sigma) = (1, 2)\sigma$ is a bijection between the set of even permutations in S_n and the set of odd permutations in S_n. As the identity permutation is even and the set of even permutations is closed under multiplication and taking inverses, the even permutations are a group.

The *alternating group* A_n is the group of even permutations of an n-element set. It has order $n!/2$.

Let G be a group and X a set. An *action* of G on X is a function that associates to each $g \in G$ and $x \in X$ an element of G, denoted gx, such that the following conditions hold:

For every $x \in X$, we have $ex = x$ (where e is the identity element of G).

For every $g, h \in G$ and $x \in X$, we have $g(hx) = (gh)x$.

In a group action, each element $g \in G$ yields a permutation of the set X, defined by sending x to gx. For if $gx = gy$, then $x = y$, so the map is one-to-one, and $gg^{-1}x = x$, so the map is onto.

For example, the symmetric group S_n acts on the set $\{1, 2, 3, \ldots, n\}$ by the action $gx = g(x)$, where $g(x)$ is the image of x under the bijection $g: \{1, 2, 3, \ldots, n\} \to \{1, 2, 3, \ldots, n\}$.

Similarly, the cyclic group $\mathbf{Z}_n$ acts on the set $\{1, 2, 3, \ldots, n\}$ by the action

$$gx = \begin{cases} g + x & \text{if } g + x \leq n \\ g + x - n & \text{if } g + x > n. \end{cases}$$

Here g denotes the equivalence class representative of $[g]$ between 1 and n.

A.5 Fields

The concept of a field arose in the study of the solvability of polynomial equations as well as in the study of properties of the real numbers and complex numbers. Heinrich M. Weber (1842–1913) gave the first modern definition of a field.

A *field* F is a set having at least two elements, with two binary operations, $+$ and $\cdot$, such that the following conditions hold:

F is an abelian group with respect to $+$.

$F - \{0\}$, where 0 is the additive identity, is an abelian group with respect to $\cdot$.

For all $x, y, z \in F$, we have

$$x \cdot (y + z) = x \cdot y + x \cdot z \quad \text{(distributive law)}.$$

Examples of fields:
The set $\mathbf{R}$ of real numbers with the usual addition and multiplication.
The set $\mathbf{Q}$ of rational numbers with the usual addition and multiplication.
The set $\mathbf{Z}_2 = \{0, 1\}$ with addition and multiplication modulo 2.

The set $\mathbf{Z}_p = \{0, 1, \ldots, p-1\}$, where p is a prime number, with addition and multiplication modulo p.

A finite field exists and is unique up to isomorphism for any prime power order. We show a construction for the order $8 = 2^3$. Start with the field $\mathbf{Z}_2 = \{0, 1\}$. Find a polynomial of degree 3 over this field that doesn't factor into polynomials of lesser degree. One choice is $f(x) = x^3 + x + 1$. We see that $f(0) = 0^3 + 0 + 1 = 1$ and $f(1) = 1^3 + 1 + 1 = 1$. So neither 0 nor 1 is a root of f. If f factored into polynomials of lesser degree, then at least one of the factors would be linear, but then 0 or 1 would be a root. Hence f doesn't factor. Next, take θ to be a root of f in the field of order 8 that we are trying to construct. Thus $f(\theta) = \theta^3 + \theta + 1 = 0$, which implies that $\theta^3 = \theta + 1$ (because the base field is $\mathbf{Z}_2$). Define the field to be the collection of polynomials of degree 2 in θ over $\mathbf{Z}_2$. The eight field elements are:

$$0, \ 1, \ \theta, \ \theta^2, \ \theta + 1, \ \theta^2 + \theta, \ \theta^2 + \theta + 1, \ \theta^2 + 1.$$

To do addition and multiplication in this field, add or multiply polynomials, reduce coefficients modulo 2, and use the identity $\theta^3 = \theta + 1$. The seven nonzero elements of the field form a cyclic group generated by θ. Successive powers of θ comprise all seven nonzero field elements.

See Appendix B for a challenge about constructing another finite field.

A.6   Vector Spaces

A vector space consists of a group of vectors defined over a field of scalars. More formally: A *vector space* V over a field F is an additive abelian group together with a rule that assigns to every $f \in F$ and $v \in V$ an element $f \cdot v \in V$ such that the following conditions hold for all $f, f_1, f_2 \in F$ and $v, v_1, v_2 \in V$:

$f \cdot (v_1 + v_2) = f \cdot v_1 + f \cdot v_2;$

$(f_1 + f_2) \cdot v = f_1 \cdot v + f_2 \cdot v;$

$f_1 \cdot (f_2 \cdot v) = (f_1 f_2) \cdot v;$

$1 \cdot v = v$, where 1 is the multiplicative identity of F.

Elements of V are called *vectors* and elements of F are called *scalars*.
Examples of vector spaces:
The group $\mathbf{R}^2$ is a vector space over the field $\mathbf{R}$.
The group $\mathbf{R}$ is a vector space over the field $\mathbf{Q}$ of rational numbers.

An important example of a vector space is F^n, the vector space of ordered n-tuples over a field F. Addition and multiplication of vectors is defined componentwise. We write an element of F as an $n \times 1$ vector. For example, with $n = 4$ and $F = \mathbf{Z}_2$, one vector is

$$\begin{bmatrix} 1 \\ 0 \\ 1 \\ 0 \end{bmatrix}.$$

Suppose that V is a vector space over F. A subset S of V *spans* V if every vector $v \in V$ can be written as a linear combination of elements of S; that is,

$$v = f_1 v_1 + \cdots + f_n v_n,$$

for some elements $v_1, \ldots, v_n$ in S and $f_1, \ldots, f_n$ in F.

A subset S of V is *linearly independent* if no element of S can be written as a linear combination of the other elements of S.

A *basis* of V is a subset of V that spans V and is linearly independent.

Theorem. *If V is a vector space, then V has a basis. Moreover, all bases of V have the same cardinality.*

The cardinality of a basis of a vector space is its *dimension*.

For example, the vector space $\mathbf{R}^2$ over $\mathbf{R}$ has dimension 2. One basis, called the *standard basis*, consists of

$$\begin{bmatrix} 1 \\ 0 \end{bmatrix} \quad \text{and} \quad \begin{bmatrix} 0 \\ 1 \end{bmatrix}.$$

A *linear transformation* from one vector space to another is given by a matrix. A *matrix* A is a rectangular array of numbers $[a_{ij}]$, where $1 \le i \le m$ and $1 \le j \le n$.

For 2×2 matrices

$$A = \begin{bmatrix} a_{11} & a_{12} \\ a_{21} & a_{22} \end{bmatrix} \quad \text{and} \quad B = \begin{bmatrix} b_{11} & b_{12} \\ b_{21} & b_{22} \end{bmatrix},$$

where the entries are arbitrary numbers, we define

$$A + B = \begin{bmatrix} a_{11} + b_{11} & a_{12} + b_{12} \\ a_{21} + b_{21} & a_{22} + b_{22} \end{bmatrix}.$$

That is, we add the corresponding entries of A and B.

We define scalar multiplication so that the result of applying the transformation A and then a multiple c is the same as applying the transformation cA. This definition amounts to multiplying each entry of A by c.

We need to define matrix multiplication. We want to define the matrix product AB so that it represents the result of applying the linear transformation B to x and y, and applying the linear transformation A to the result. We write x and y as a vector

$$\begin{bmatrix} x \\ y \end{bmatrix}.$$

We set

$$\begin{bmatrix} a_{11} & a_{12} \\ a_{21} & a_{22} \end{bmatrix} \begin{bmatrix} x \\ y \end{bmatrix} = \begin{bmatrix} a_{11}x + a_{12}y \\ a_{21}x + a_{22}y \end{bmatrix},$$

and

$$\begin{bmatrix} a_{11} & a_{12} \\ a_{21} & a_{22} \end{bmatrix} \begin{bmatrix} b_{11} & b_{12} \\ b_{21} & b_{22} \end{bmatrix} \begin{bmatrix} x \\ y \end{bmatrix} = \begin{bmatrix} a_{11} & a_{12} \\ a_{21} & a_{22} \end{bmatrix} \begin{bmatrix} b_{11}x + b_{12}y \\ b_{21}x + b_{22}y \end{bmatrix}$$

$$= \begin{bmatrix} (a_{11}b_{11} + a_{12}b_{21})x + (a_{11}b_{12} + a_{12}b_{22})y \\ (a_{21}b_{11} + a_{22}b_{21})x + (a_{21}b_{12} + a_{22}b_{22})y \end{bmatrix}$$

$$= \begin{bmatrix} a_{11}b_{11} + a_{12}b_{21} & a_{11}b_{12} + a_{12}b_{22} \\ a_{21}b_{11} + a_{22}b_{21} & a_{21}b_{12} + a_{22}b_{22} \end{bmatrix} \begin{bmatrix} x \\ y \end{bmatrix}.$$

Therefore, we define

$$\begin{bmatrix} a_{11} & a_{12} \\ a_{21} & a_{22} \end{bmatrix} \begin{bmatrix} b_{11} & b_{12} \\ b_{21} & b_{22} \end{bmatrix} = \begin{bmatrix} a_{11}b_{11} + a_{12}b_{21} & a_{11}b_{12} + a_{12}b_{22} \\ a_{21}b_{11} + a_{22}b_{21} & a_{21}b_{12} + a_{22}b_{22} \end{bmatrix}.$$

We call the ij entry of the product the *dot product* of the ith row vector of A and the jth column vector of B.

Matrix addition is defined similarly for any two matrices of the same dimensions, and matrix multiplication is defined similarly for any two matrices in which the number of columns of the first matrix is the same as the number of rows of the second matrix.

We can use matrices to solve systems of linear equations. For example, we can write the system

$$3x + 4y + 5z = -154$$

$$-x + 10z = 0$$

$$3x + 7y + 12z = -385$$

as

$$\begin{bmatrix} 3 & 4 & 5 \\ -1 & 0 & 10 \\ 3 & 7 & 12 \end{bmatrix} \begin{bmatrix} x \\ y \\ z \end{bmatrix} = \begin{bmatrix} 60 \\ -91 \\ 6 \end{bmatrix}.$$

The matrix

$$\begin{bmatrix} x \\ y \\ z \end{bmatrix}$$

is a vector. Call it v. Denote the 3×3 matrix by A and the vector of constants on the right side by c. The system is

$$Av = c.$$

We can solve this system by multiplying by the inverse of A. Suppose that A^{-1} is the inverse of A (with respect to matrix multiplication). Then

$$v = A^{-1}c.$$

When is the matrix

$$A = \begin{bmatrix} a_{11} & a_{12} \\ a_{21} & a_{22} \end{bmatrix}$$

invertible? Solving for x and y in the corresponding two-equation system $Ax = c$, we find that

$$x = \frac{c_1 a_{22} - c_2 a_{12}}{a_{11} a_{22} - a_{12} a_{21}} \quad \text{and} \quad y = \frac{c_2 a_{11} - c_1 a_{21}}{a_{11} a_{22} - a_{12} a_{21}}.$$

The quantity $a_{11} a_{22} - a_{12} a_{21}$ is called the *determinant* of the matrix. The system is solvable if and only if the determinant is nonzero.

The *determinant* of an $n \times n$ matrix $A = [a_{i,j}]$ is defined as

$$\det A = \sum_\sigma \text{sgn}(\sigma) a_{1,\sigma(1)} a_{2,\sigma(2)} \cdots a_{n,\sigma(n)},$$

where the sum is over all $n!$ permutations σ of the set $\{1, 2, \ldots, n\}$ and $\text{sgn}(\sigma)$ is the sign of σ, i.e., $+1$ if σ is an even permutation and -1 if σ is an odd permutation.

Let's look at a matrix as a geometric transformation. Consider a rotation of the Cartesian coordinate system by θ radians in the counterclockwise direction about the origin. We can find the matrix that performs this transformation. Trigonometry shows that

$$\begin{bmatrix} 1 \\ 0 \end{bmatrix}$$

is rotated to

$$\begin{bmatrix} \cos \theta \\ \sin \theta \end{bmatrix},$$

and

$$\begin{bmatrix} 0 \\ 1 \end{bmatrix}$$

is rotated to

$$\begin{bmatrix} -\sin \theta \\ \cos \theta \end{bmatrix}.$$

It follows from the definition of matrix multiplication that the rotation matrix is given by

$$R_\theta = \begin{bmatrix} \cos \theta & -\sin \theta \\ \sin \theta & \cos \theta \end{bmatrix}.$$

The definitions make such computations transparent.

ℬ

Eye-Opening Explorations

The only way to learn mathematics is to do mathematics.

—PAUL HALMOS (1916–2006)

In this appendix, we pose some problems related to topics discussed in the book. Can you solve these problems?

B.1 Problems

1. Recall the Lemniscate of Chapter 1. A Cartesian equation for this curve is

$$(x^2 + y^2)^2 = x^2 - y^2.$$

Suppose that x and y are integers considered modulo p, where p is an odd prime. Use a computer algebra system to count the number of ordered pairs (x, y) that satisfy the lemniscate equation modulo p. Conjecture a formula for the number of solutions in terms of p.

2. Recall the discussion of a googol in Centillion in Chapter 1. What is the smallest number of pennies such that the number of subsets is greater than a googol?

3. Recall the properties of complex numbers and determinants discussed in Chapter 1.

   (a) Prove that two triangles whose vertices in the complex plane are α, β, γ and α', β', γ' are similar if and only if

$$\begin{vmatrix} \alpha & \beta & \gamma \\ \alpha' & \beta' & \gamma' \\ 1 & 1 & 1 \end{vmatrix} = 0.$$

   (b) Prove that the complex numbers α, β, and γ are the vertices of an equilateral triangle if and only if

$$\alpha + \beta\omega + \gamma\omega^2 = 0,$$

   where

$$\omega = -\frac{1}{2} + \frac{\sqrt{3}}{2}i$$

   is a cube root of unity.

4. In this problem, we look at a flat version of the Square Pyramidal Square Number of Chapter 2. Find an integer greater than 1 that is both a square number and a triangular number.

5. In Bulging Hyperspheres of Chapter 2, we saw that a "small" hypersphere of radius $\sqrt{d}-1$ bulges outside a d-dimensional hypercube of side 4 when $d > 9$. What happens to the ratio of the volume of this hypersphere to the volume of the hypercube as d increases? Use the formula from Volume of a Ball in Chapter 3.

6. Recall the Two-Colored Graph in Chapter 2. Draw a three-coloring of the edges of the complete graph on 16 vertices with the property that there is no triangle all of whose edges are the same color.

   Refer to the discussion of Hypercube in Chapter 2.

   Prove that a three-coloring of the edges of a complete graph on 17 vertices must contain a triangle with all three edges the same color.

   Use the pigeonhole principle (see page 66).

   These results are prominent in the area of combinatorics known as Ramsey theory.

7. Recall the Hypercube of Chapter 2. If we define a hypercube in 5 dimensions, how many vertices does it have? How many neighboring vertices does each vertex have? How many edges are there?

8. Recall the Squaring Map of Chapter 2. For $n = 10^6$, find the number of components of the graph and the size of a largest attractor. You may need a computer.

9. Recall the Riemann Sphere of Chapter 2. What action on the Riemann sphere is caused by the mapping $z \mapsto 1/z$ in the complex plane? What geometric relationship do the numbers z and $-1/\overline{z}$ have?

10. Recall the Heronian triangles of Chapter 3. Find two incongruent triangles with integer side lengths, having the same integer area and perimeter. You may need a computer.

11. Use the technique for finding the area of a triangle given in Heron's Formula and Heronian Triangles in Chapter 3 to find the volume of a tetrahedron in terms of its side lengths.

12. Recall the product formula for $\sin x$ given in Product for Pi in Chapter 3. Prove the infinite product formula for the hyperbolic sine function:

$$\prod_{n=1}^{\infty} \frac{n^2 + 1}{n^2} = \frac{\sinh \pi}{\pi}.$$

13. Use mathematical induction to prove Cassini's identity, from Fibonacci Numbers and Pi in Chapter 3:

$$F_n^2 - F_{n+1}F_{n-1} = (-1)^{n+1}, \quad n \geq 1.$$

Use this and the formula for the tangent of a difference to prove the partial sums formula

$$\sum_{n=1}^{k} \tan^{-1} \frac{1}{F_{2n+1}} = \frac{\pi}{4} - \frac{1}{F_{2k+2}}.$$

14. Recall the definition of the Fibonacci sequence $\{F_n\}$ given in Fibonacci Numbers and Pi in Chapter 3. Find its generating function. Also, find the generating function for the sequence of fifth powers of the Fibonacci numbers, $\{F_n^5\}$.

15. Recall The Smallest Taxicab Number of Chapter 3. Find the smallest positive integer that is the sum of two fourth powers in more than one way.

16. In the discussion of The Zeta Function and Bernoulli Numbers in Chapter 3, we said that we could use Bernoulli numbers to find

$$\sum_{m=1}^{\infty} \frac{1}{m^4}.$$

Fill in the details of this calculation.

17. Recall the Riemann zeta function from Chapter 3. Prove that

$$\zeta(-n) = -\frac{B_{n+1}}{n+1}, \quad n \geq 1.$$

In particular, this means that $\zeta(-n) = 0$ when n is an even positive integer.

18. Give a counting proof of the recurrence formula for the number of Rook paths in Chapter 3:

$$r(0,0) = 1, \; r(0,1) = 1, \; r(1,0) = 1, \; r(1,1) = 2;$$

$$r(m,n) = 2r(m-1,n) + 2r(m,n-1) - 3r(m-1,n-1),$$

$$m \geq 2 \text{ or } n \geq 2.$$

19. Recall the theorem A Square inside Every Triangle from Chapter 4. Describe how to perform the construction of a square inside a triangle using straightedge and compass.

20. Recall from Polynomial Symmetries in Chapter 4 that given a finite group, there is a polynomial whose symmetries are that group. Find such a polynomial for the dihedral group D_6. See Figure A.1.

21. Recall the definition of Kings and Serfs in a tournament from Chapter 4. A vertex that reaches every other vertex in one step is called an *Emperor*. Prove that

   (a) A tournament with no Emperor has at least three Kings.
   (b) A tournament on $n > 4$ vertices can have any number of Kings between 1 and n except 2.

22. How many permutations of the set $\{1, 2, 3, \ldots, 16\}$ have no increasing subsequence of length five and no decreasing subsequence of length five? Recall the discussion after the Erdős–Szekeres theorem in Chapter 4.

23. Use Minkowski's theorem from Chapter 4 to prove that, for any real numbers a, b, c, d with

$$\Delta = \begin{vmatrix} a & b \\ c & d \end{vmatrix} \neq 0,$$

there exist integers x and y, not both 0, such that

$$|ax + by| \leq \sqrt{\Delta} \quad \text{and} \quad |cx + dy| \leq \sqrt{\Delta}.$$

24. Recall the lemniscate graph from Chapter 4. Find the area it encloses.

25. The substitution $r = \sin t$ changes the integral giving the length of the lemniscate curve from

$$L = 4 \int_0^{\pi/2} \frac{dt}{\sqrt{1 + \sin^2 t}}$$

to

$$L = 4 \int_0^1 \frac{dr}{\sqrt{1 - r^4}}.$$

Use this integral to show

$$L = 4 \left(1 + \frac{1}{2 \cdot 5} + \frac{1 \cdot 3}{2 \cdot 4 \cdot 9} + \frac{1 \cdot 3 \cdot 5}{2 \cdot 4 \cdot 6 \cdot 13} + \cdots \right).$$

The last number in each denominator is twice the previous number plus 1.

26. Investigate the Fibonacci sequence $\{F_n\}$ modulo powers of 2. Can you conjecture a formula for the length of the period of the sequence $\{F_n \bmod 2^m\}$, where m is a positive integer? Do the same for the sequence $\{F_n \bmod 5^m\}$. This problem comes from the discussion of Integer Triangles in Chapter 5.

27. Find a formula for the number of odd entries in the nth row of Pascal's triangle.

Recall Odd Binomial Coefficients in Chapter 5.

28. Use the method of constructing Perfect Error-Correcting Codes from Chapter 5 to construct a binary code of length eight, consisting of 16 code words at distance at least four apart. Start with a cyclic graph on four vertices. What code do you get when you delete one coordinate from the length eight code?

29. Recall the problem Making a Million in Chapter 6. How many ways are there to make a million dollars if we also use $2 bills?

30. Prove the claim about the expected number of steps in the one-dimensional gobbling algorithm described in Two-Dimensional Gobbling Algorithm in Chapter 7.

31. Show how to construct a field of nine elements.

Recall the explanation of constructing finite fields in Appendix A. Start with the field $\mathbf{Z}_3$. Find a polynomial of degree 2 that doesn't factor over this field.

Recall the discussion of Projective Plane in Chapter 2. Explain how to use the nine point field to construct a projective plane with 91 points and 91 lines.

B.2   Solutions

1. If $p \not\equiv 1 \pmod 8$, then the curve contains p integer points, while if $p \equiv 1 \pmod 8$, then it contains $p - 4$ integer points. We find a parameterization of the lemniscate with rational functions. Recalling the technique of Diophantus from Squares in Arithmetic Progression in Chapter 5, we use the rational parameterization of the unit circle

$$\cos t = \frac{2v}{1 + v^2}, \quad \sin t = \frac{1 - v^2}{1 + v^2}, \quad v \in \mathbf{Q}.$$

From this, we obtain a rational parameterization for the lemniscate:

$$x = \frac{v + v^3}{1 + v^4}, \quad y = \frac{v - v^3}{1 + v^4}, \quad v \in \mathbf{Q}.$$

If we plot this curve in the plane, where v is a real variable, we see that the entire lemniscate is generated with each point occurring exactly once. If v is in the field $\{0, 1, 2, \ldots, p - 1\}$, then we have to interpret what the division in the formula means and what happens if a denominators is 0. If the denominators are nonzero, then we interpret division as multiplying by a multiplicative inverse.

When can the denominators $1 + v^4$ be 0 modulo p? The nonzero elements modulo p, where p is a prime, form a cyclic group of order $p - 1$, with multiplication modulo p. Let g be a generator of the group. Then $-1 = g^{(p-1)/2}$, and we see that -1 has a fourth root if and only if $(p - 1)/4$ is an integer, i.e., $p \equiv 1 \pmod 8$. In this case, the four solutions to $1 + v^4 = 0$ are $g^{(p-1)/8}$, $g^{3(p-1)/8}$, $g^{5(p-1)/8}$, and $g^{7(p-1)/8}$. For any prime p, the defined points in the parameterization are distinct. This accounts for p points on the lemniscate if $p \not\equiv 1 \pmod 8$ and $p - 4$ points if $p \equiv 1 \pmod 8$. To show that the values are distinct, assume that

$$\frac{v_1(1 + v_1^2)}{1 + v_1^4} = \frac{v_2(1 + v_2^2)}{1 + v_2^4}, \quad \frac{v_1(1 - v_1)}{1 + v_1^4} = \frac{v_2(1 - v_2)}{1 + v_2^4}.$$

Dividing the first expression by the second and simplifying, we obtain

$$\frac{1 + v_1^2}{1 - v_1^2} = \frac{1 + v_2^2}{1 - v_2^2},$$

or

$$-1 + \frac{2}{1 - v_1^2} = -1 + \frac{2}{1 - v_2^2},$$

which implies that $v_1^2 = v_2^2$. Substituting into the original equations yields $v_1 = v_2$.

2. A computer check shows that $n = 333$ is the smallest integer such that

$$2^n > 10^{100}.$$

3. (a) Two triangles are similar if and only if one can be dilated, rotated, and translated to the position of the other. The first row of the given matrix is a linear combination of the second and third rows (and the determinant is 0) when

$$(\alpha, \beta, \gamma) = a(\alpha', \beta', \gamma') + b(1, 1, 1).$$

The multiplier a does the dilation and rotation (a complex number can accomplish this), while $b(1, 1, 1)$ is a translation in the direction of the complex number b.

(b) The complex numbers α, β, and γ are the vertices of an equilateral triangle if and only if

$$\omega(\gamma - \beta) = \alpha - \gamma.$$

To see this, sketch the vectors $\gamma - \beta$ and $\alpha - \gamma$, and see what happens when the first vector is rotated $120°$ counterclockwise (done by ω). Using the relation $1 + \omega + \omega^2 = 0$, we have

$$\alpha + \beta\omega + \gamma\omega^2 = 0.$$

4. It is easy to check that 36 is the smallest such number. We have

$$36 = 6^2 = 1 + 2 + 3 + \cdots + 8.$$

All square triangular numbers are given by the recurrence formula

$$a_0 = 0, \ a_1 = 1, \quad a_n = 6a_{n-1} - a_{n-2}, \ n \geq 2.$$

Thus $\{a_n\} = \{0, 1, 6, 35, 204, \ldots\}$, and a_n^2 is both square and triangular.

5. The ratio of the volume of the hypersphere of radius $\sqrt{d} - 1$ to the volume of the hypercube of side 4 is

$$\frac{\pi^{d/2}(\sqrt{d} - 1)^d}{(d/2)!4^d}.$$

As d increases, the volume decreases close to 0 and then takes off toward infinity.

6. The picture below shows a graph representing one of the color classes. This graph can be made from the five-dimensional hypercube with each point also joined to its diagonal opposite. In binary representation this means joining each binary string to the strings that differ from it in one coordinate and to its complementary string.

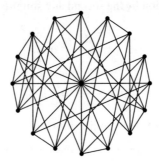

The three color classes must be put together to make the complete graph on sixteen vertices. I leave this as an exercise.

Now we show that every three-coloring of the edges of the complete graph on seventeen vertices must contain a monochromatic triangle. Choose a vertex. There are sixteen edges emanating from it. Since there are three colors available for the edges, it follows by the pigeonhole principle that at least six of them are the same color. Without loss of generality, suppose that six edges emanating from the vertex are blue. If any edge joining two endpoints of these edges is blue, then there is a blue triangle and we are done. So suppose that all the edges on the complete graph formed by the six endpoints are red or green. Then, by the version of Ramsey's theorem mentioned in Two-Colored Graph, there is a red triangle or a green triangle and we are done.

7. From the model of a hypercube in dimension n as the set of binary strings of length n, we see that in dimension 5 there are $2^5 = 32$ vertices. The neighbors of a vertex are the strings that differ from it in exactly one coordinate. Since there are five choices for the coordinate, each vertex has five neighbors. Multiplying the number of vertices by the number of neighbors per vertex, we get the total number of edges counted twice (from the perspective of both endpoints). Hence, the hypercube has $32 \cdot 5/2 = 80$ edges.

An n-dimensional hypercube has a vertex set consisting of all binary n-tuples. Two vertices are joined by an edge if they differ in exactly one coordinate. For $0 \le k \le n$, a k-dimensional face of an n-dimensional hypercube is a subset of k coordinates and the corresponding edge connections that form a k-dimensional hypercube. The number of k-dimensional faces of an n-dimensional hypercube is

$$\binom{n}{k}2^{n-k}, \quad 0 \le k \le n.$$

The reason is that there are $\binom{n}{k}$ choices for the k coordinates that form the k-dimensional hypercube. The other $n - k$ coordinates can be either 0 or 1.

For example, the 4-dimensional hypercube has $2^4 = 16$ vertices, $\binom{4}{1}2^3 = 32$ edges, $\binom{4}{2}2^2 = 24$ faces, $\binom{4}{3}2 = 8$ three-dimensional faces, and $\binom{4}{4} = 1$ four-dimensional face (the whole hypercube). Try this calculation for the 3-dimensional cube.

8. A simple algorithm finds the number of components and the size of a largest attractor in the squaring map modulo n. Start with n singleton sets $\{0\}, \{1\}, \{2\}, \ldots, \{n-1\}$. For k from 0 to $n - 1$, concatenate the set containing k and the set containing $k^2 \bmod n$. Then the sets are the components of the graph. To find the size of the attractor in each set, choose an element of a set and compute its successive squares modulo n. When the process goes into a loop, count the number of steps in the loop. Applying this algorithm (using a computer) when $n = 10^6$, we find that there are fourteen components and a largest attractor has size 2500.

9. The Riemann sphere is rotated $180°$ about the real (x-) axis. To see this, show that the mapping $z \mapsto 1/z$ induces the mapping $(x, y, z) \mapsto (x, -y, -z)$. The numbers z and $-1/\bar{z}$ are antipodal points on the Riemann sphere.

10. A simple computer program finds the smallest two such triangles: $\{17, 25, 28\}$ and $\{20, 21, 29\}$, both with perimeter 70 and area 210. You can find this solution by hand with a little educated guessing. Heron's formula says that the area of a triangle with sides $\{a, b, c\}$ is $\sqrt{s(s-a)(s-b)(s-c)}$, where $s = (a+b+c)/2$. Setting $a' = s-a$, $b' = s-b$, and $c' = s-c$, the formula becomes $\sqrt{sa'b'c'}$. Noting that $a' + b' + c' = s$, the problem requires finding two triples $\{a', b', c'\}$ with equal products and equal sums. Experimenting with prime factors 2, 3, 5, and 7, we find two such triples: $\{2 \cdot 3^2, 2 \cdot 5, 7\}$

and $\{3 \cdot 5, 2 \cdot 7, 2 \cdot 3\}$, both with sum $5 \cdot 7$; and the product $sa'b'c'$ is a perfect square. The triples $\{a', b', c'\}$ determine the triples $\{a, b, c\}$.

11. Assume that the tetrahedron has sides a, b, c, a', b', c', with a' opposite to a, b' opposite to b, and c' opposite to c. Suppose that a is represented by the vector $\mathbf{a}$, etc. The volume of the tetrahedron is

$$V = \frac{1}{6} |\det M|,$$

where M is the 3×3 matrix whose rows are $\mathbf{a}$, $\mathbf{b}$, and $\mathbf{c}$. Since the transpose matrix M^t has the same determinant as M, we have

$$36V^2 = \det M \det M^t = \det MM^t = \begin{vmatrix} a^2 & \mathbf{a} \cdot \mathbf{b} & \mathbf{a} \cdot \mathbf{c} \\ \mathbf{a} \cdot \mathbf{b} & b^2 & \mathbf{b} \cdot \mathbf{c} \\ \mathbf{a} \cdot \mathbf{c} & \mathbf{b} \cdot \mathbf{c} & c^2 \end{vmatrix}.$$

Writing the dot products in terms of the side lengths, we obtain

$$36V^2 = \begin{vmatrix} a^2 & (a^2 + b^2 - c'^2)/2 & (a^2 + c^2 - b'^2)/2 \\ (a^2 + b^2 - c'^2)/2 & b^2 & (b^2 + c^2 - a'^2)/2 \\ (a^2 + c^2 - b'^2)/2 & (b^2 + c^2 - a'^2)/2 & c^2 \end{vmatrix}$$

$$288V^2 = \begin{vmatrix} 2a^2 & a^2 + b^2 - c'^2 & a^2 + c^2 - b'^2 \\ a^2 + b^2 - c'^2 & 2b^2 & b^2 + c^2 - a'^2 \\ a^2 + c^2 - b'^2 & b^2 + c^2 - a'^2 & 2c^2 \end{vmatrix}.$$

A computer algebra system can help to calculate the determinant. It turns out that

$$144V^2 = a^2 b^2 a'^2 + a^2 b^2 b'^2 + a^2 c^2 a'^2 + a^2 c^2 c'^2$$
$$+ b^2 c^2 b'^2 + b^2 c^2 c'^2 + a^2 a'^2 b'^2 + a^2 a'^2 c'^2$$
$$+ b^2 a'^2 b'^2 + b^2 b'^2 c'^2 + c^2 b'^2 c'^2 + c^2 a'^2 c'^2$$
$$- a^2 b^2 c'^2 - a^2 c^2 b'^2 - b^2 c^2 a'^2 - a'^2 b'^2 c'^2$$
$$- a^2 a'^4 - b^2 b'^4 - c^2 c'^4 - a'^2 a^4 - b'^2 b^4 - c'^2 c^4.$$

The first twelve terms are products of squares of triples that constitute neither a triangle nor three edges with a common vertex. The next four terms are products of squares of triples that form a triangle. The last six terms are products of squares of sides and fourth powers of their opposite sides.

12. By definition,

$$\sinh z = \frac{e^{iz} - e^{-iz}}{2}.$$

The zeros of $\sinh z$ are $z = in\pi$, where n is an integer. Hence, an infinite product expansion of this function is

$$\sinh z = z \prod_{n=1}^{\infty} \left(1 + \frac{z^2}{n^2 \pi^2}\right).$$

Letting $z = \pi$, we obtain the infinite product formula.

13. Cassini's identity holds for $n = 1$, since

$$F_1^2 - F_2 F_0 = 1 - 1 \cdot 0 = 1 = (-1)^2.$$

Assume that it holds for n. Then

$$\begin{aligned}
F_{n+1}^2 - F_{n+2} F_n &= F_{n+1}^2 - (F_n + F_{n+1})F_n \\
&= F_{n+1}(F_{n+1} - F_n) - F_n^2 \\
&= F_{n+1} F_{n-1} - F_n^2 \\
&= -(-1)^{n+1} \\
&= (-1)^{n+2},
\end{aligned}$$

and we see that the identity holds for $n + 1$. Therefore, by mathematical induction, Cassini's identity holds for all $n \geq 1$.

From the formula for the tangent of a difference and Cassini's identity,

$$\begin{aligned}
\tan^{-1} \frac{1}{F_{2n}} - \tan^{-1} \frac{1}{F_{2n+2}} &= \tan^{-1} \frac{(1/F_{2n}) - (1/F_{2n+2})}{1 + (1/F_{2n})(1/F_{2n+2})} \\
&= \tan^{-1} \frac{F_{2n+2} - F_{2n}}{F_{2n} F_{2n+2} + 1} \\
&= \tan^{-1} \frac{F_{2n+1}}{F_{2n+1}^2} \\
&= \tan^{-1} \frac{1}{F_{2n+1}}.
\end{aligned}$$

It follows that the partial sums are telescoping series:

$$\begin{aligned}
\sum_{n=1}^{k} \tan^{-1} \frac{1}{F_{2n+1}} &= \sum_{n=1}^{k} \left(\tan^{-1} \frac{1}{F_{2n}} - \tan^{-1} \frac{1}{F_{2n+2}} \right) \\
&= \tan^{-1} \frac{1}{F_2} - \tan^{-1} \frac{1}{F_{2k+2}} \\
&= \frac{\pi}{4} - \tan^{-1} \frac{1}{F_{2k+2}}.
\end{aligned}$$

14. To find the generating function for the Fibonacci sequence, we use the well-known formula

$$F_n = \frac{\phi^n - \hat{\phi}^n}{\sqrt{5}}, \quad n \geq 0,$$

where $\phi = (1 + \sqrt{5})/2$ and $\hat{\phi} = (1 - \sqrt{5})/2$. Then

$$\begin{aligned}
\sum_{n=0}^{\infty} F_n x^n &= \frac{1}{\sqrt{5}} \sum_{n=0}^{\infty} \left(\phi^n - \hat{\phi}^n \right) x^n = \frac{1}{\sqrt{5}} \left(\frac{1}{1 - \phi x} - \frac{1}{1 - \hat{\phi} x} \right) \\
&= \frac{x}{1 - x - x^2}.
\end{aligned}$$

The series converges for $|\phi x| < 1$.

We find the generating function for the fifth powers of the Fibonacci numbers similarly. From the binomial theorem and the fact that $\phi\hat\phi = -1$, we have

$$F_n^5 = \frac{1}{(\sqrt5)^5}\left(\phi^n - \hat\phi^n\right)^5$$

$$= \frac{1}{25\sqrt5}\left(\phi^{5n} - 5\phi^{4n}\hat\phi^n + 10\phi^{3n}\hat\phi^{2n} - 10\phi^{2n}\hat\phi^{3n} + 5\phi^n\hat\phi^{4n} - \hat\phi^{5n}\right)$$

$$= \frac{1}{25\sqrt5}\left((\phi^5)^n - 5(-\phi^3)^n + 10\phi^n - 10\hat\phi^n + 5(-\hat\phi^3)^n - (\hat\phi^5)^n\right).$$

Thus, the generating function is

$$\frac{1}{25\sqrt5}\left(\frac{1}{1-\phi^5 x} - \frac{5}{1+\phi^3 x} + \frac{10}{1-\phi x} - \frac{10}{1-\hat\phi x} + \frac{5}{1+\hat\phi^3 x} - \frac{1}{1-\hat\phi^5 x}\right).$$

Combining the outer two fractions, then the next outer two, and finally the inner two, we obtain

$$\frac{1}{5}\left(\frac{x}{1-11x-x^2} + \frac{2x}{1+4x-x^2} + \frac{2x}{1-x-x^2}\right).$$

We use the fact that the Lucas numbers, L_n, defined by the recurrence relation $L_0 = 2$, $L_1 = 1$, and $L_n = L_{n-1} + L_{n-2}$, for $n \geq 2$, are given by $L_n = \phi^n + \hat\phi^n$, for $n \geq 0$. The generating function is valid for $|\phi^5 x| < 1$.

15. Leonhard Euler found the solution

$$635318657 = 59^4 + 158^4 = 133^4 + 134^4.$$

A straightforward way to find it using a computer is to generate a list of the first 200 fourth powers, then form a list of sums of two terms from the first list. The solution will appear twice on the second list.

16. The formula

$$\zeta(2n) = (-1)^{n+1}\frac{(2\pi)^{2n}}{2(2n)!}B_{2n}$$

with $n = 2$ yields

$$\zeta(4) = -\frac{(2\pi)^4}{2(4)!}B_4.$$

Since $B_4 = -1/30$, we obtain

$$\zeta(4) = \frac{\pi^4}{90}.$$

17. Applying the formula

$$\zeta(s) = 2^s\pi^{s-1}\sin\left(\frac{\pi s}{2}\right)\Gamma(1-s)\zeta(1-s), \quad \Re s < 0,$$

for $s = -n$, we have

$$\zeta(-n) = 2^{-n}\pi^{-n-1}\sin\left(\frac{\pi(-n)}{2}\right)\Gamma(n+1)\zeta(n+1).$$

If n is even, then the sine factor is 0, so $\zeta(-n) = 0$. If n is odd, then we may use the formulas

$$\zeta(2k) = (-1)^{k+1} \frac{(2\pi)^{2k}}{2(2k)!} B_{2k}, \quad k \geq 1,$$

and

$$\Gamma(n+1) = n!$$

to obtain

$$\zeta(-n) = -\frac{B_{n+1}}{n+1}.$$

18. From the definition of $r(m, n)$, we have

$$2r(m-1, n) + 2r(m, n-1) - 3r(m-1, n-1)$$

$$= r(m-1, n) + r(m-1, n) + r(m, n-1)$$

$$\quad + r(m, n-1) - 3(m-1, n-1)$$

$$= \sum_{a=0}^{m-1} r(a, n) + \sum_{b=0}^{n-1} r(m-1, b) + \sum_{b=0}^{n-1} r(m, b) + \sum_{a=0}^{m-1} r(a, n-1)$$

$$\quad - 3(m-1, n-1)$$

$$= \sum_{a=0}^{m-1} r(a, n) + \sum_{b=0}^{n-1} r(m, b) + \sum_{b=0}^{n-1} r(m-1, b) + \sum_{a=0}^{m-1} r(a, n-1)$$

$$\quad - 3(m-1, n-1)$$

$$= \sum_{a=0}^{m-1} r(a, n) + \sum_{b=0}^{n-1} r(m, b) + 3(m-1, n-1) - 3(m-1, n-1)$$

$$= r(m, n).$$

19. Here is a diagram of the construction.

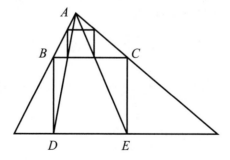

The construction of a square ($BCED$) by straightedge and compass is well known. Draw lines from D and E to A. The points where they intersect BC give the side of the square to be constructed. A similar triangles argument shows that the rectangle inscribed in $\triangle ABC$ is a square.

20. One such polynomial is

$$x_1 x_2 + x_2 x_3 + x_3 x_4 + x_4 x_5 + x_5 x_6 + x_1 x_6.$$

21. (a) Suppose that the tournament has no Emperor. Let v be a King. Let A be the set
of vertices to which v is directed and B the remaining set of vertices. Since v is a
King, v can reach any vertex in B in two steps. Since v is not an Emperor, B is not
empty. Why is A non-empty? Let v_A be a King in the tournament restricted to A,
and v_B a King in the tournament restricted to B. Then v_A and v_B are also Kings of
the given tournament, so the tournament has at least three Kings.

(b) If the tournament has an Emperor, then there is exactly one King, and we are done.
By part (a), if there is no Emperor then there are at least three Kings. So we must
show that any number of Kings from 3 to n is possible. Suppose that $3 \leq k \leq n$.
By the proof of Maurer's theorem, a tournament on k vertices has the property that
not every vertex is a King with probability at most

$$\binom{k}{2}\left(\frac{3}{4}\right)^{k-2}.$$

This is less than 1 for $k \geq 21$. Hence, there exists a tournament on k vertices in
which every vertex is a King when $k \geq 21$. As an exercise, you can construct
tournaments with this property for $3 \leq k \leq 20$. Now, let T be a tournament on k
vertices in which every vertex is a King. Draw a directed edge from every vertex of
T to $n - k$ other vertices. Join them to each other with edges directed in any way.
This tournament has exactly k Kings.

22. According to the hook length formula, the number of standard fillings of a 4×4 grid
with the numbers 1 through 16 is

$$\frac{16!}{1 \cdot 2 \cdot 2 \cdot 3 \cdot 3 \cdot 3 \cdot 4 \cdot 4 \cdot 4 \cdot 4 \cdot 5 \cdot 5 \cdot 5 \cdot 6 \cdot 6 \cdot 7} = 24024.$$

The number of permutations of the integers 1 through 16 that do not contain an increasing subsequence of length five or a decreasing subsequence of length five is

$$24024^2 = 577152576.$$

23. Let K be the set of ordered pairs (x, y) of real numbers such that

$$-\sqrt{\Delta} \leq ax + by \leq \sqrt{\Delta}$$
$$-\sqrt{\Delta} \leq cx + dy \leq \sqrt{\Delta}.$$

The transformation $(x, y) \mapsto (ax + by, cx + dy)$ is linear with determinant Δ. It
follows that the area of K is 4. If the area of K were greater than 4, then by Minkowski's
theorem K would contain a lattice point other than $(0, 0)$. However, the same conclusion
holds for K since it is a closed set. Why does it still hold?

24. The area bounded by a curve given by parametric equations $x(t)$ and $y(t)$, where $\alpha \leq t \leq \beta$, is

$$\int_\alpha^\beta y(t)x'(t)\,dt.$$

For the lemniscate, the integral is

$$4\int_{\pi/2}^0 \frac{4\cos t \sin^2 t\,(5 + \cos(2t))}{(-3 + \cos(2t))^3}\,dt.$$

An antiderivative is

$$F(t) = \frac{-4\sin^3 t}{(-3 + \cos(2t))^2},$$

and so, by the Fundamental Theorem of Calculus, the area is

$$4(F(0) - F(\pi/2)) = 1.$$

25. Isaac Newton (1643–1727) showed how to extend the binomial theorem to all real exponents by the series formula

$$(1 + x)^\alpha = \sum_{n=0}^\infty \binom{\alpha}{n} x^n, \quad |x| < 1,$$

where

$$\binom{\alpha}{n} = \frac{\prod_{i=0}^{n-1}(\alpha - i)}{n!}, \quad n \geq 1, \quad \binom{\alpha}{0} = 1.$$

The integrand is

$$\frac{1}{\sqrt{1 - r^4}} = 1 + \frac{1}{2}r^4 + \frac{1 \cdot 3}{2 \cdot 2 \cdot 2!}r^8 + \frac{1 \cdot 3 \cdot 5}{2 \cdot 2 \cdot 2 \cdot 3!}r^{12} + \cdots.$$

Integrating this over $0 \leq r \leq 1$ yields the desired series.

26. The period of $\{F_n \mod 2^m\}$ is $3 \cdot 2^{m-1}$. The period of $\{F_n \mod 5^m\}$ is $4 \cdot 5^m$. No general formula is known for the period of the sequence $\{F_n \mod m\}$ where $m \geq 2$.

27. Recall that n dominates k means that the binary expansion of n has a 1 in every position where the binary expansion of k has a 1. Suppose that the binary expansion of n has $d(n)$ 1s. Then n dominates exactly $2^{d(n)}$ numbers, as this is the number of subsets of a set of size $d(n)$. Therefore, there are $2^{d(n)}$ odd entries in the nth row of Pascal's triangle.

28. The first four coordinates of the code represent a collection of vertices of the cyclic graph of length four. There are $2^4 = 16$ such subsets. The last four coordinates comprise the indicator vector of the set of vertices nonadjacent to the first set. The resulting code has distance four. Deleting a coordinate results in a code of length seven and distance three, consisting of 16 code words. It is a Hamming code.

29. The number of ways is

$$40125046347199679029952380920610239599328534571 30267501$$

$$\doteq 4 \times 10^{54}.$$

30. Let $e(n)$ be the expected number of steps in the gobbling algorithm starting with n. The first number chosen is either 1 (with probability $1/n$) or not 1 (with probability $(n-1)/n$). This yields the recurrence formula

$$e(1) = 1; \quad e(n) = \frac{1}{n}(e(n-1)+1) + \frac{n-1}{n}e(n-1) = e(n-1) + \frac{1}{n}, \quad n \geq 2.$$

It follows that

$$e(n) = 1 + \frac{1}{2} + \frac{1}{3} + \cdots + \frac{1}{n}.$$

31. The polynomial $f(x) = x^2 + x + 2$ does not factor over $\mathbf{Z}_3 = \{0, 1, 2\}$. To see this, try 0, 1, and 2, and check that you do not get 0. If f factored, then it would have two linear factors and hence would have roots in $\mathbf{Z}_3$. Let θ be a root of f. A field F of nine elements is obtained by taking powers of θ together with 0. Thus, the elements of the field are

$$0, \ 1, \ \theta, \ \theta^2 = 2\theta + 1, \ \theta^3 = 2\theta + 2, \ \theta^4 = 2, \ \theta^5 = 2\theta, \ \theta^6 = \theta + 2, \ \theta^7 = \theta + 1.$$

To construct a projective plane of order nine, let the points be the ordered pairs (x, y), where $x, y \in F$, together with the ideal points $m \in F$ and ∞. This accounts for $9^2 + 9 + 1 = 91$ points. The lines are of three types: sets of points (x, y) that satisfy an equation $y = mx + b$, where $m, b \in F$, together with m; sets of points (a, y), where $a \in F$, together with ∞; and the ideal line, consisting of the points $m \in F$ and ∞. This accounts for $9^2 + 9 + 1 = 91$ lines. As an exercise, check that each point is on ten lines, each line contains ten points, every two points determine a unique line, and every two lines intersect in a unique point.

There exist three other projective planes of order nine that do not arise from a field. See [7].

Bibliography

[1] M. Aigner and G. M. Ziegler. *Proofs From THE BOOK*. Springer-Verlag, New York, third edition, 2004.

[2] C. Alsini and R. B. Nelsen. A visual proof of the Erdős–Mordell inequality. *Forum Geometricorum*, 7:99–102, 2007.

[3] G. E. Andrews and K. Eriksson. *Integer Partitions*. Cambridge University Press, Cambridge, 2004.

[4] W. S. Anglin. The square pyramid puzzle. *The American Mathematical Monthly*, 97(2):120–124, 1990.

[5] R. Bacher and S. Eliahou. Extremal matrices without constant 2-squares. *Journal of Combinatorics*, 1(1):77–100, 2010.

[6] L. W. Beineke and R. J. Wilson, editors. *Graph Connections*. Clarendon Press, Oxford, 1997.

[7] M. K. Bennett. *Affine and Projective Geometry*. Wiley, New York, 1995.

[8] D. Bindner and M. Erickson. Alcuin's sequence. *The American Mathematical Monthly*, to appear.

[9] G. Boros and V. H. Moll. *Irresistible Integrals: Symbolics, Analysis and Experiments in the Evaluation of Integrals*. Cambridge University Press, Cambridge, 2004.

[10] E. Brown. Three Fermat trails to elliptic curves. *The College Mathematics Journal*, 31(3):162–172, 2000.

[11] S. A. Burr. On moduli for which the Fibonacci sequence contains a complete system of residues. *The Fibonacci Quarterly*, 9:497–504, 1971.

[12] K. Ciesielski. *Set Theory for the Working Mathematician*, volume 39 of *London Mathematical Society Student Texts*. Cambridge University Press, Cambridge, 1997.

[13] J. H. Conway, H. Burgiel, and C. Goodman-Strauss. *The Symmetries of Things*. A. K. Peters, Wellesley, 2008.

[14] J. H. Conway and N. J. A. Sloane. *Sphere Packings, Lattices, and Groups*. Springer–Verlag, New York, third edition, 1999.

[15] M. Erickson. *Pearls of Discrete Mathematics*. Chapman & Hall/CRC Press, Boca Raton, 2009.

[16] M. Erickson, S. Fernando, and K. Tran. Enumerating rook and queen paths. *Bulletin of the Institute of Combinatorics and Its Applications*, 60(37-48), 2010.

[17] M. J. Erickson. *Introduction to Combinatorics*. Wiley, New York, 1996.

[18] M. J. Erickson and J. Flowers. *Principles of Mathematical Problem Solving*. Prentice Hall, Upper Saddle River, 1999.

[19] S. Glaz and J. Growney, editors. *Strange Attractors: Poems of Love and Mathematics*. A. K. Peters, New York, 2009.

[20] A. M. Gleason. Angle trisection, the heptagon, and the triskaidecagon. *The American Mathematical Monthly*, 95(3):185–194, 1988.

[21] R. L. Graham, D. E. Knuth, and O. Patashnik. *Concrete Mathematics: A Foundation for Computer Science*. Addison-Wesley, Reading, MA, second edition, 1994.

[22] R. L. Graham, B. L. Rothschild, and J. H. Spencer. *Ramsey Theory*. Wiley, New York, second edition, 1990.

[23] G. Hardy and E. Wright. *An Introduction to the Theory of Numbers*. Clarendon Press, Oxford, fifth edition, 1989.

[24] J. Hemmeter. On an iteration diagram. *Congressium Numeratium*, 60:59–66, 1987.

[25] I. N. Herstein. *Abstract Algebra*. Prentice Hall, Upper Saddle River, third edition, 1996.

[26] R. Honsberger. *Mathematical Gems I*. Mathematical Association of America, Washington, DC, 1973.

[27] D. L. Johnson. *Presentations of Groups*. Cambridge University Press, New York, 1990.

[28] T. Koshy. *Fibonacci and Lucas Numbers with Applications*. Wiley-Interscience, New York, 2001.

[29] C. F. Laywine and G. L. Mullen. *Discrete Mathematics Using Latin Squares*. Wiley-Interscience, New York, 1998.

[30] M. Levi. *The Mathematical Mechanic: Using Physical Reasoning to Solve Problems*. Princeton University Press, Princeton, 2009.

[31] B. Lindstrom and H.-O. Zetterstom. Borromean circles are impossible. *The American Mathematical Monthly*, 98(4):340–341, 1991.

[32] J. H. van Lint and R. M. Wilson. *A Course in Combinatorics*. Cambridge University Press, Cambridge, 1992.

[33] W. Miller. The maximum order of an element of a finite symmetric group. *The American Mathematical Monthly*, 94(6):497–506, 1987.

[34] G. Minton. Three approaches to a sequence problem. *Mathematics Magazine*, 84(1):33–37, 2011.

[35] E. Nagel and J. R. Newman. *Gödel's Proof*. New York University Press, New York, 2001.

[36] P. J. Nahin. *An Imaginary Tale: The Story of $\sqrt{-1}$*. Princeton University Press, Princeton, 1998.

[37] I. Niven, H. Zuckerman, and H. Montgomery. *An Introduction to the Theory of Numbers*. Wiley, New York, fifth edition, 1991.

[38] H. Noon and G. van Brummelen. The non-attacking queens game. *The College Mathematics Journal*, 37(3):223–227, 2006.

[39] C. D. Olds, A. Lax, and G. Davidoff. *The Geometry of Numbers*. The Mathematical Association of America, Washington, DC, 2000.

[40] R. W. Owens. An algorithm to solve the Frobenius problem. *Mathematics Magazine*, 76(4):264–275, 2003.

[41] V. Pambuccian. The Erdős–Mordell inequality is equivalent to non-positive curvature. *Journal of Geometry*, 88:134–139, 2008.

[42] M. Petkovsek, H. Wilf, and D. Zeilberger. *A=B*. AK Peters, New York, 1996.

[43] C. Petzold. *Code: The Hidden Language of Computer Hardware and Software*. Microsoft Press, Redmond, 1999.

[44] T. N. Phan and N. M. Mach. Proof for a conjecture on general means. *Journal of Inequalities in Pure and Applied Mathematics*, 9(3), 2008.

[45] V. Pless. *Introduction to the Theory of Error-Correcting Codes*. Wiley, New York, third edition, 1998.

[46] S. Schwartzman. *The Words of Mathematics: An Etymological Dictionary of Mathematical Terms Used in English*. Mathematical Association of America, Washington, DC, 1994.

[47] H. Shniad. On the convexity of mean value functions. *Bulletin of the American Mathematical Society*, 54:770–776, 1948.

[48] D. Smith, M. Eggen, and R. St. Andre. *A Transition to Advanced Mathematics*. Brooks/Cole, Chicago, seventh edition, 2009.

[49] A. Stacey and P. Weidl. The existence of exactly m-coloured complete subgraphs. *Journal of Combinatorial Theory, Series B*, 75:1–18, 1999.

[50] P. D. Straffin. *Game Theory and Strategy*. Mathematical Association of America, Washington, DC, 1993.

[51] L. W. Tu. *An Introduction to Manifolds*. Springer, New York, 2008.

[52] H. Walser. *The Golden Section*. Mathematical Association of America, Washington, DC, 2001.

[53] L. C. Washington. *Elliptic Curves: Number Theory and Cryptography*. Chapman & Hall/CRC Press, Boca Raton, 2003.

[54] D. B. West. *Introduction to Graph Theory*. Prentice Hall, Upper Saddle River, 1995.

[55] P. Yiu. Heronian triangles are lattice triangles. *The American Mathematical Monthly*, 108:261–263, 2001.

Index

About the Author

Martin Erickson was born in Detroit, MI in 1963. He graduated with High Honors from the University of Michigan in 1985 and received his Ph.D. at the University of Michigan in 1987. He is a professor of mathematics at Truman State University. He has written several acclaimed mathematics books, including *Aha! Solutions* (MAA) and *Introduction to Number Theory* (with Anthony Vazzana, CRC Press). He is a member of the Mathematical Association of America and the American Mathematical Society.